The Energy Crisis

Other Johns Hopkins Books of Related Interest

Ethical Land Use: Principles of Policy and Planning
TIMOTHY BEATLEY

Getting the Word Out in the Fight to Save the Earth
RICHARD BEAMISH

Greenways for America
CHARLES E. LITTLE

Losing Asia: Modernization and the Culture of Development
BRET WALLACH

Natural Hazards: An Integrative Framework for Research and Planning
RISA I. PALM

Planning for Earthquakes: Risk, Politics, and Policy
PHILIP R. BERKE AND TIMOTHY BEATLEY

Saving America's Countryside: A Guide to Rural Conservation
SAMUEL N. STOKES, A. ELIZABETH WATSON, *et alia*

Soil Conservation in the United States: Policy and Planning
FREDERICK R. STEINER

Superfund: The Political Economy of Environmental Risk
JOHN A. HIRD

Water Resources Management: In Search of an Environmental Ethic
DAVID LEWIS FELDMAN

The Energy Crisis

Unresolved Issues and Enduring Legacies

Edited by David Lewis Feldman

THE JOHNS HOPKINS UNIVERSITY PRESS   BALTIMORE AND LONDON

© 1996 The Johns Hopkins University Press
All rights reserved. Published 1996
Printed in the United States of America on acid-free paper
05   04   03   02   01   00   99   98   97   96   5   4   3   2   1

The Johns Hopkins University Press
2715 North Charles Street, Baltimore, Maryland 21218–4319
The Johns Hopkins Press Ltd., London

Published in cooperation with the
Center for American Places, Harrisonburg, Virginia.

Library of Congress Cataloging-in-Publication Data
will be found at the end of this book.

A catalog record for this book is available from
the British Library.

ISBN 0-8018-5361-3

To my wife, Debbie

Contents

Foreword

Milton Russell

Q41

For the United States, the energy shock of 1973–74 was a pivotal event of the last quarter of the twentieth century. Coming on the heels of the Vietnam War, it exposed our economic vulnerability to forces outside our borders and further eroded confidence in our ability to control events. It weakened the hold of technological optimism, especially after the failed efforts at "energy independence." It ushered in a sense of limits—a sense that, indeed, perpetual expansion and ever-increasing exploitation of natural resources could not prevail. Beyond its direct economic impacts, it changed the way people thought about the future—or, more precisely, it made thinking about the future salient in ways it had never been before.

There is another dimension to the effects of the oil shock that is of particular relevance to many readers of this book. The response to the shock mobilized and trained a small army of analysts to understand its implications and develop the tools required to affect its consequences. Under the forced-draft pressures of economic disarray, political anxiety, and international tension, analysts were pressed to predict energy prices, to estimate production, consumption, and import levels, and to quantify the effects of myriad possible policy actions on a multitude of dimensions. As some of the contributors to this book testify, the analytical response on the whole was exemplary. The basic tenets of economics provided useful insights into demand and supply response, and the tools of policy analysis modeled the results of alternative government policies with precision sufficient to inform decisions. Participants in the process recall the heady swirl of nightless days and weekendless weeks as they simultaneously built the tools and models and did analysis in real time as events unfolded. Mistakes were made, but on the whole the analytical establishment got the answers roughly right.

From the efficacy of price decontrol to the predictions that international oil prices would not soar, from the economic and technical futility of "energy independence" to the prospects for natural gas, the analysts were consistently closer to the mark than the popular opinion that was driving national policy, as subsequent events bore out.

When it came to the crucial decisions taken, however, the political process was governed by considerations other than economics. Fear of the political effects of energy price increases, efforts to use the energy crisis to mold new international alliances, xenophobic pressures, and suspicion of and hostility toward the oil industry all led to policies that analysis at the time marked as flawed, and that hindsight has proven as such. Still, the analysis leaned against the wind, provided the basis for political leaders to resist more egregious actions, and built the basis for reversing failed policies as time and experience allowed political pressures to dissipate.

While the basic economic and policy analysis expertise was available and burgeoned quickly, knowledge of the structure and institutions of the energy industries and of world oil markets was in short supply. In Herbert Stein's memorable phrase, experts were anyone who knew there were "42 gallons of oil in a barrel; and that Khadafi was a person and Abu Dhabi was a place." There were a few souls lost in the bowels of the Department of Interior who had been quietly gathering data on oil markets, but for the most part the government was not prepared. Fortuitously, some academic expertise was available, and it was quickly marshaled. It is useful to note from where it came.

The Ford Foundation, in its Energy Policy Project completed in 1974, was a major contributor to building both a shelf of useful research and a coterie of people who had been thinking about these issues. The regulatory travails of the natural gas industry had led a few persons to do policy research that spilled over to oil and coal markets. The controversy over oil import controls had a similar effect. Perhaps the most important intellectual force, however, came from Resources for the Future (RFF), which under the leadership of Joe Fisher, Sam Schurr, and Hans Landsberg had nurtured research and researchers on energy issues for over a decade before the crisis erupted. Indeed, on the eve of the energy shock, RFF published Morrie Adelman's crucial book, *The World Petroleum Industry* (1972).

On an illustrative, personal note, in 1973, well before the Arab oil embargo, RFF enlisted Doug Bohi and myself to do a study of world oil markets as they affected the United States. After the crisis intervened, RFF then graciously allowed us to interrupt this work to do a study for the

Congress, one of the first to attempt to explain oil price behavior in the OPEC era (*Oil Imports and Energy Security: Analysis of the Current Situation and Future Prospects,* 1974). The RFF work was published in 1975 (*U.S. Energy Policy: Alternatives for Security*) and, after a delay occasioned by my service at the Council of Economic Advisers, in 1978 (*Limiting Oil Imports*).

Building on the base of available academic expertise, and with an assured market for analysis, persons with new skills and enthusiasms joined those with institutional and industry knowledge to bring forth more orderly and insightful ways of understanding energy markets and the forces that play among them. The methodologies developed in this effort have since been applied elsewhere, and in the ensuing twenty years, careers molded in this earlier period have enriched a broad swath of U.S. public policy.

This record offers some important lessons for our time. It illustrates the importance of the U.S. tradition of the government being open to short-term service by academics and others. Almost unique in the world, the United States brings private-sector expertise and leadership into government and has the capacity to augment its ranks with appointees whose careers are grounded elsewhere. Recent developments, including Draconian postemployment conflict-of-interest regulations and the unseemly pillorying of presidential appointees in the Senate confirmation process, threaten one of the major strengths of our society in responding to future crises.

Further, the preparation for the energy shock illustrates the strength and resiliency provided by dispersed sources of research funding. Government funding, especially for policy-related research, is driven toward problems that are known and approaches that are accepted. Without private-sector funding, especially with the situation in today's universities, research that does not meet government's perceived needs is stymied. Now, with even government funding in short supply, the need for increased support for creative ideas is reaching a crisis point. The patient nurturing of work that prepares the country for challenges beyond the horizon is simply not taking place at the level required.

Finally, this record offers a sobering caution to America's foundations, major sponsors of policy and other research in the past. In a quest for relevance and control, foundations appear to be narrowing their focus, moving more directly into social action enterprises, and looking for short-term, tangible results. As they do so, they shorten the time horizons for the work they sponsor and replicate the funding available elsewhere.

The consequence is that funding is restricted for the innovative and far-reaching research programs that foundations are uniquely positioned to foster. Increasingly, research proposals to foundations are taking on the same format and employ the same criteria for success—narrow focus, clear methodology, sharply defined problem, predictable results, tight deadlines—as do proposals to other funders such as the National Science Foundation. While understandable, this trend weakens the intellectual resource base the country will have to deal with future developments. As a revealing example of different foundation strategies in the past, consider the major post-energy-shock analyses that were designed to garner lessons to prepare for the road ahead. The Andrew W. Mellon Foundation underwrote Resources for the Future's *Energy in America's Future* (Sam H. Schurr et al., 1979) and the Ford Foundation—sponsored *Energy: The Next Twenty Years* (Hans Landsberg et al., 1979), which, along with Robert Stobaugh and Daniel Yergin's *Energy Future* (1979), provided the best summaries of where the country stood and where it could go as it came to terms with adjustment to the events of 1973–74.

The purpose of the symposium that led to this book was to examine the twenty years since the first energy shock to see what its effects actually were, as a way of opening a window to the future. It brought together some of the key players of twenty years ago for a period of disciplined reminiscence. More important, it joined them with other analysts and observers to explore what we might expect for the future.

The result is a book that works on many levels. It captures a slice of history of how ideas were formed and conveyed during a period of stress and change. It offers insights into how government policies were formed, and suggests caution about how past mistakes may be repeated if events conspire again. It dissects, from different perspectives, how the energy shock was propagated through the economy, and its proximate and more distant consequences. It demonstrates where vulnerabilities still lie, and describes plausible actions to reduce them. It looks beyond the present to suggest what might be done now to smooth the transition to a sustainable energy path for the future. Beyond all this, it instructs by example in showing how disciplined minds have approached one of the more daunting intellectual challenges presented to public policy in the recent past.

Milton Russell

Joint Institute for Energy and Environment
The University of Tennessee–Knoxville

Preface and Acknowledgments

This book stems from a symposium entitled "Twenty Years after the Energy Shock—How Far Have We Come? Where Are We Headed?" which was held at the University of Tennessee–Knoxville on April 19, 1994, to commemorate the university's bicentennial. The event was organized by the Energy, Environment, and Resources Center (EERC), a multidisciplinary research center at the university that explores critical energy, technology, and natural resource issues and the Joint Institute for Energy and Environment (JIEE). Speakers at this daylong symposium included authorities from Resources for the Future, the Electric Power Research Institute, Harvard University, the World Bank, the National Academy of Sciences, Worldwatch Institute, the Tennessee Valley Authority, the U.S. Department of Energy, the U.S. Environmental Protection Agency, the *Washington Post*, Massachusetts Institute of Technology, Louisiana State University, the American Petroleum Institute, Oak Ridge National Laboratory, Energy Market and Policy Analysis, Inc., and the University of Tennessee.

An edited book, especially the product of a symposium, is a collective effort. When the University of Tennessee's Bicentennial Committee made funds available to the EERC to produce a public, commemorative event, several people helped shape the energy symposium.

Jack Barkenbus, the EERC's director, provided matching financial support for the symposium and for the production of this book. His encouragement of this final product was invaluable.

Along with Jack, Milton Russell, director of the JIEE—an institute composed of Oak Ridge National Laboratory, the Tennessee Valley Authority, and the University of Tennessee—was instrumental in the symposium. Milt conceived the theme, "What have we learned since the 1970s oil price shock?" He also served on the symposium's advisory board and was a symposium co-chair. Milt's tireless effort in rounding up additional sponsorship and participants was essential to the event's success. I might add that he contributed additional financial support on behalf of the JIEE.

Intellectually, several of Milt's peers at Resources for the Future, especially Hans Landsberg and Doug Bohi, inspired many of the ideas that found their way into the symposium.

Steven R. Brechin, of the Department of Sociology and the Center for Energy and Environmental Studies at Princeton University, was an excellent peer reviewer for the Johns Hopkins University Press. His comments were most valuable as I prepared the final manuscript.

My special thanks go to the other sponsors of the symposium who made possible the production of this book: Len Coburn of the U.S. Department of Energy; George Hidy of the Electric Power Research Institute; and Michael Canes of the American Petroleum Institute.

I would also like to thank the other members of the symposium advisory committee: Bob Bohm of the Department of Economics and EERC at the University of Tennessee; Mary English of EERC; and Bob Shelton, director of the Energy Division of Oak Ridge National Laboratory.

The logistical and clerical support of Debbie Bower, Tina Cordy, Sherry Estep, and Gail Farris of the University of Tennessee, and Hemal Tailor, a UT student, is also acknowledged with great appreciation. I would also like to thank Elithe Carnes for her effort in preparing the index.

Last but not least, to the authors who toiled to produce the manuscripts that became this book—my thanks for your efforts.

The Energy Crisis

1. Revisiting the Energy Crisis

Unresolved Impacts, Enduring Legacies

DAVID LEWIS FELDMAN

More than twenty years have passed since the Arab oil embargo of 1973–74 disrupted supplies of imported petroleum, drove up the price of gasoline, forced Americans to wait in line at service stations, and generated fear that a little-known cartel called OPEC, the Organization of Petroleum Exporting Countries, might bring the U.S. economy to its knees. The embargo shook public confidence in government and exacerbated distrust of big business. Most of all, it dramatized America's dependence on imported oil and the consequences of its insatiable appetite for energy.

The "energy crisis," as it came to be called, has three enduring legacies. First, it exposed how normal market behavior can be adversely affected by political instability and subsequent sudden price increases. In short, cheap foreign oil can be vulnerable to sudden, unanticipated disruption. Second, it showed how well-intentioned policies designed to preserve wilderness from rampant energy development, to reduce air and water pollution, and to foster coal mine safety could affect the price and availability of energy.

Finally, it initiated a public policy debate over the appropriate role of government and markets in ensuring a stable, economical, and environmentally sustainable energy supply. This debate hinges on a central question: Should government directly subsidize energy research and development (R&D), restrict energy imports, and impose conservation programs on businesses and individuals through taxes and regulation; or should these activities be encouraged by unleashing market forces that are responsive to supply, demand, and price? While the U.S. government no longer controls the interstate transit of natural gas or sets gasoline prices, debate over government's role in promoting energy R&D, in regulating fuel efficiency, and in controlling the environmental externalities of energy remains very much alive. It remains alive in part because of lingering concern that we as a society may have prematurely declared "victory" in the battle to address our energy problems squarely and comprehensively. Questions pertaining to the relationship between foreign political instability and energy supply, government's role in promoting energy R&D, and the interrelationships between energy and environment remain unanswered.

One need cite only three recent examples to illustrate the continuing legacy of the events of 1973–74: the Clinton administration's ill-fated "BTU tax," designed to reduce the federal deficit while also weaning consumers away from fossil fuels; the Climate Change Action Plan and its attempt to couple voluntary energy-efficiency and conservation efforts with federal investments; and the current study by the U.S. Department of Energy (DOE) of the future of the national laboratories and their role in energy R&D.[1] These issues also show the need for continued discussion of an integrated, as opposed to fragmented, national energy strategy.

This book addresses four issues related to this legacy:
—*lessons learned* from the energy shocks of the 1970s about the causes and consequences of government intervention in energy markets and the effects of sudden adversity upon the stability of governments and markets
—*enduring impacts* of the energy crisis, including the continuing difficulty in predicting energy futures and prescribing appropriate actions, and the reverberating effects of the energy shocks on society
—*unresolved issues,* including prospects for environmentally sustainable and publicly acceptable energy sources that can be brought into widespread use
—*the future of energy,* including the prognosis for R&D, new technologies, and the appropriate balance between governments and markets in making choices

Lessons Learned

What have we learned over the past twenty years about reducing dependence on embargo-vulnerable liquid fuels? What factors influence energy markets and their imperfections? Does government intervention cause, resolve, or merely aggravate these imperfections? When the experts whose views are collected here convened at the University of Tennessee in April 1994 to discuss these matters, they concurred on two lessons: (1) the energy crisis did not begin with the Arab oil embargo (in fact, the embargo was, on the whole, a trivial event); and (2) the crisis was probably resolved by a combination of market forces and government intervention.

Unintended Causes

One popular line of reasoning contends that the energy crisis was caused by a string of decisions implemented by policymakers many years before the oil embargo. These decisions were designed to ensure that Americans would have a cheap, reliable energy supply and a cleaner environment. In reality, according to Resources for the Future's Doug Bohi and Joel Darmstadter (chap. 2), these decisions exacerbated energy shortages in four ways.

First, environmental regulations introduced during the late 1960s and early 1970s, particularly the Clean Air and Mine Safety and Health Acts, made coal use more expensive and less attractive. They did this by imposing worker health and safety regulations on an industry suffering from low productivity and by placing stringent emission controls on consumers (particularly industries and utilities).

Second, domestic oil drilling was encouraged by wellhead taxes imposed on some domestic producers, as well as by subsidies on oil imports. Particularly noteworthy was the mandatory oil import program, begun in 1959, which imposed limits on oil imports, encouraged depletion of domestic sources—and thus, ironically, increased oil imports by raising demand—and distorted oil production decisions. Ironically, the oil import program acted as a tax by channeling demand toward imports (Bohi and Russell, 1978).

Third, the selling price of natural gas, which had been federally regulated two decades before the energy crisis, was held below its market value and replacement cost, reducing incentives for producers to search for, and open, new gas fields. Interstate price controls increased gas consumption and distorted the allocation of gas to consumers at the very same time that growing demand for gas caused the price to rise in unregulated intrastate markets. These markets absorbed new supplies and forced consumers in regulated interstate markets to switch to oil. This in turn helped accelerate the demand for oil in the years leading up to the embargo of 1973.

Finally, regulations on coal produced a synergistic effect on oil and gas production. At precisely the time that these regulations began to take effect in the late 1960s and early 1970s, nationwide shortages of natural gas emerged. Because of the costs of regulation, the coal industry, having suffered from decades of decline exacerbated by low prices, was unable to meet growing energy demands and to relieve the pressure on oil supplies.

According to Bohi and Darmstadter's reasoning, government intervention aggravated the embargo by inhibiting the development of alternative
sources of fuel able to provide readily available substitutes when they were
needed. As a result, public policies helped transform a relatively minor
supply interruption into a full-fledged economic crisis.

There is a second line of reasoning pertaining to the origins of the energy
crisis. While not disputing these seminal causes (i.e., market distortions
caused by public policies that encouraged consumption of oil and reduced
production of coal and gas), this line of reasoning diverges slightly when it
comes to ascribing immediate blame. As described by the *Washington Post*'s
John Berry (chap. 6), some of the conditions that exacerbated the oil price
shocks of 1973–74 were unavoidable consequences of public policies that
were perceived as both beneficial and desirable. For example, nuclear energy costs increased dramatically in the early 1970s, in part because of
environmental and safety regulations imposed on utilities. Regardless of
their adverse price impact, these regulations were intended to bolster public
confidence in a relatively nascent industry that, even before Three Mile
Island, was perceived as laden with potential risks from operator error or
mechanical failure. Likewise, few complained when coal prices increased as
a result of federal regulations, because at the time, the latter were viewed as
desirable.

Most important, however, when the price shocks finally occurred as a
result of the 1973–74 embargo, Berry argues, a number of events arose that
limited the federal government's ability to respond swiftly and effectively.
Persistent inflation generated by the recycling of financial surpluses in oil-
exporting countries, coupled with three recessions in the span of eight
years, fiscal policy disarray, and a new wave of foreign business competition
(which benefited from increased U.S. energy costs), made the execution of a
"recovery" policy exceptionally difficult. As Berry notes, the crisis fed antipathy toward the domestic oil industry, making it unpopular for policymakers to urge patience with "market corrections," on the one hand, or
with modest government intervention to wean the country away from foreign oil dependence, on the other. Moreover, the advent of a debt crisis in
less developed countries, exacerbated by higher energy prices, placed severe
constraints on American policymakers' ability to impose unilateral decisions beneficial to us, because they could be potentially adverse to some of
our third world allies (e.g., Saudi Arabia, Kuwait).

Purdue's George Horwich (chap. 5), while not disagreeing with this interpretation of events following the embargo, offers a slightly different ascrip-

tion of blame. According to Horwich, many economists and decision makers were also reluctant to end price and allocation controls following the embargo for fear that they might actually compound supply disruptions. These proponents argued that in the absence of controls, further shortages could drastically raise prices, unleash massive consumer spending, and create windfall profits. If Horwich is correct, the inertia of policy precedent may be a stronger factor in explaining the severity of the energy crisis than conventional analyses recognize.

Crisis Resolution

The predominant view among the contributors to this volume, reflecting a perspective common in mainstream economic theory, is that rising energy prices, coupled with changes in international energy markets, created greater flexibility in supply by terminating long-term relationships and contracts. In short, when the producing countries gained control of energy supplies from the large oil companies, a more flexible market was created that encouraged energy exploration, production, and distribution. In the long term, as Louisiana State University's Allan Pulsipher suggests (chap. 4), higher prices compelled consumers to conserve energy and make more prudent choices about energy consumption. Prior to 1973 we imported oil because it was cheap to do so. As Pulsipher notes, we continue to do so today for the same reason, but with one difference: the oil market is now more responsive to demand and to spot availability.

However, despite continued dependence on foreign oil, the higher prices charged for imported oil beginning in the mid-1970s have permitted some beneficial market "correction." According to Pulsipher, higher prices have made available so-called spot-market diversification within the petroleum industry, easing shortages of oil. Price increases have also provided consumers with incentives to conserve energy, and producers with incentives to explore for new sources.

While this view of how the immediate crisis ended is favored by most of the contributors to this volume, it is not universally endorsed. A few contend that this mainstream interpretation of the easing of the energy crisis, while largely valid, pays too little regard to exogenous factors that also helped ease the crisis. These factors include the introduction of new technologies, the imposition of energy-efficiency regulations, and the emergence of national-interest rivalries among OPEC members that divided the cartel against itself. This minority viewpoint, exemplified in somewhat different ways by Oak Ridge National Laboratory's David Greene (chap. 10) and Worldwatch Institute's Christopher Flavin (chap. 15), contends that it would be reckless to disregard the view that government intervention following the embargo helped to correct market imperfections caused in part by shortsighted decision-maker horizons.

According to Greene, while government decisions must shoulder some of the blame for the 1973 energy shock, nevertheless government provided, and continues to provide, the only coherent decision-making framework to mobilize the authority needed to overcome market imperfections. This authority is exemplified by decisions to impose fuel-economy regulations on motor vehicles. Fuel-economy regulations inspired by the energy crisis have led to a 50 percent increase in the fuel efficiency of new trucks and a doubling of efficiency for cars. More important, notes Greene, had these changes awaited market "demand" to be brought about, they might never have occurred.

Vehicle manufacturers have little incentive to produce fuel-efficient cars. Consumers do not buy cars by calculating the "discount rate" of future gasoline purchases. When fuel-efficiency standards initially led to more expensive vehicles, however, consumers still bought them. In the long run, manufacturers ended up being better-off, because their products successfully competed with fuel-stingy imports.

For his part, Flavin comes to similar conclusions but through very different methods. While clearly not optimistic about our ability to survive another liquid-fuel embargo, Flavin suggests that one legacy of the 1973–74 crisis was to bolster those energy analyses, both privately supported and government-sponsored, that acknowledged the need for more flexible energy planning scenarios. New technologies and energy sources, different mixes of fuels, and changing consumer patterns are difficult to predict yet important to apprehend in making energy policy. Assuming that societies need an unbiased clearinghouse to compile and disseminate such scenarios, and some means to prod the introduction of new technologies that slip through the cracks of the marketplace, then one task of government is to provide such activities, Flavin argues.

What Greene and Flavin imply is something that other observers of energy policy also noted after 1973–74. Once the energy crisis had erupted, Americans expected their government to take decisive action to resolve a shortage of petroleum that caused considerable economic inconvenience and hardship. Follow-up governmental responses, designed to encourage the development of alternative sources of energy and to prevent such a crisis from recurring, may not have been economically rational, but they may have been politically prudent. This principle goes a long way toward explaining policymaker response to the embargo. In the mid-1970s American voters continued to demand cheap and abundant energy. Economic logic may have dictated that prices for all energy sources should have been permitted to rise unrestricted. However, the political logic of trying to build a minimum winning coalition in Congress may explain such controversial—and later ill-fated—decisions as development of a federally subsidized synthetic fuels industry, and Congress' initial refusal to permit market forces alone to shape energy prices (Uslaner, 1989). In a democracy based on the accountability of elected officials to voters, consumer patience sometimes takes a back seat to voter impulse.

If this political explanation of governmental response to the energy crisis is true, then there may be some consolation for free-market proponents in the fact that public clamor for an activist government heavily committed to regulation historically fluctuates with the perceived effectiveness of that regulation. As DOE's Peter Fox-Penner suggests (chap. 17), when government intervention is perceived as generating more problems than it solves, the tide of public opinion tends to swing toward less intervention and more market discipline.

Moreover, as Denny Ellerman of the Massachusetts Institute of Technology points out (chap. 3), the current tide of opinion against energy regulation would seem to confirm Fox-Penner's observations. Faith in the ability of government to continue to do the things that Greene and Flavin would have it do has been significantly eroded. Many energy policies have failed to fulfill their promises (e.g., energy independence and cheap renewable technologies). Today the balance between public- and private-sector involvement in energy has shifted to the point where public-sector means of influencing energy policy, while robust, are probably less pronounced than in the 1970s. Imposing taxes, controlling access to public lands, promoting R&D, signaling potential supply threats through use of a strategic oil reserve, and using government as a "bully pulpit" are the principal means of federal government influence. States, meanwhile, have some control over energy policy through rate setting and other regulatory mechanisms imposed through public service commissions. In such policy areas as utility demand-side management, these are potentially significant powers. Some would argue that more government restraint has led to more abundant energy supplies, fewer hidden subsidies for special interests, and less waste on frivolous, low-payoff R&D experiments.

If Americans perceive the status quo as "not broken," perhaps there is little incentive to seek a public-sector fix. The contributors to this volume would at least agree that such perceptions tend to be cyclical. However, given government's present role as a "sidelines" player, the question must be asked: If another energy crisis were to occur and the public were to clamor for something to be done to fix the problem and resolve the crisis, would government be ready to respond in a meaningful fashion? To some degree the answer depends upon how much policymakers have learned about the behavior of energy markets, a subject discussed in the next section.

Enduring Impacts: The Continuing Difficulty of Making Predictions

Have energy analysts improved their ability to predict the behavior of energy markets and the impact of policy decisions on them? Can analysts accurately assess the continuing impacts of the energy crisis on the U.S. economy? The contributors to this volume differ sharply on these issues. However, they largely concur upon one implication that ties both issues together: sound decision making requires sound information and analysis. Without it, decision makers have little hope of making the right decisions. With it, public policies may still fail if the information and analysis are unclear, incoherent, inconsistent, or poorly communicated by the analyst. The need for sound decision making and the difficulty of finding good information remain problems today.

These problems also lie at the heart of radically divergent explanations for the continuing effects of the energy crisis on American society. As noted earlier, John Berry, whose commentary serves to tie together the first and second topics of this volume, is mostly concerned with the political damage wrought by the embargo. According to Berry, the energy crisis gave birth to, and exacerbated, many structural problems that chronically afflict America's economy.

While Berry suggests that we adopt a less sanguine view of the need for rapid, effective government response when such crises occur, an unspoken question lies at the center of his commentary: Why has it taken so long for us to figure out how to do so? Finding an answer to that question requires that we better understand the work of the energy policy analysts who work in universities, national laboratories, government agencies, and nongovernmental organizations. Today, decision makers continue to rely on these analysts; and as in the 1970s, the product of their labors, while sometimes right and sometimes wrong, is almost always influential.

Contrary to popular misconception, analytical forecasting as a tool of policy analysis is not new. Interest in predicting the future is driven by the desire to "aid analysts and decision makers in seeing problems early" (Starling, 1979: 75). As in other policy areas, the penchant for forecasting in energy policy is intended to anticipate energy problems while we can still presumably do something about them.

Harvard University's William Hogan (chap. 7) contends that there are four possible results regarding the insights offered by these analyses: (1) good insights acquired and successfully communicated to policymakers; (2) good insights acquired but not communicated, representing a "first-order" failure of policy analysis; (3) a failed prediction rigorously followed, leading to a failed policy; and (4) a failed policy resulting from a good model used incorrectly.

Category 1 is exemplified by four policy successes: the rejection of "Project Independence" (the Nixon administration's attempt to achieve absolute independence from foreign oil at any cost); the successful estimation of supply and demand elasticities in energy markets and their implications for the relationship between energy use and economic growth; analyses of the effects of price and allocation controls on energy markets; and analyses of competitive forces on natural gas and electricity transportation markets despite the apparent presence of natural monopolies.

The latter led to the deregulation of natural gas through removal of price controls, in phases, during the 1970s and early 1980s. This decision, while made at some political peril—cheap gas, if short in supply, was politically popular—was clearly the right thing to do.[2] Economic models rightly predicted that removing controls would hasten exploration, production, and distribution, and would ensure a more stable, abundant supply. Conversely, Hogan feels that the establishment of the Strategic Petroleum Reserve (SPR) was a bad decision based on incorrect forecasting. To Bohi and Darmstadter the SPR was the result of a wise and prudent decision; its use in a test phase prior to the Persian Gulf War should be applauded, and it has had the effect of stabilizing prices. Hogan, on the other hand, believes that the SPR has been unnecessary.

The subject of forecasting raises an interesting question, however: If bad predictions lead to bad policy, then do good predictions always produce better policies? Not necessarily. Here, according to Hogan, empiricism is a better guide than logic. Predictions are only as good as their interpretation. Just as a bad prediction followed closely can produce bad policy, so too can a good prediction not adhered to closely enough. An example, Hogan notes, is the failure to pay sufficient attention to the short-lived threat from OPEC. OPEC's decline and the triumph of market pricing were missed by many forecasters and policymakers alike, Hogan says, because they failed to understand the behavior of cartels. Whereas Hogan commends economic analysis for furthering this understanding, Allan Pulsipher implies that political causes, in the form of nationalism, may provide a far better explanation for OPEC's decline.

The American Petroleum Institute's Michael Canes (chap. 8) largely agrees with Hogan's analysis, but he focuses his commentary on whether or not energy predictions have led to better policy. The answer, Canes contends, depends in part upon how one defines *better*. While the federal government may have abandoned the notion of mandating reduced dependence on foreign energy sources, many states continue to subsidize alternative energy sources.

Glenn Schleede, of Energy Market and Policy Analysis, Inc., suggests that such actions by government should be viewed as a means to strengthen the power of entrenched interest groups, senior government executives, or members of Congress. Thus, they will continue to be promoted regardless of their economic merit. On the other hand, to the extent that public skepticism toward continued support for such policies remains high, then politics—not economics—will serve to check the growth of these sometimes misguided overtures.

If Hogan, Canes, and Schleede show the limits of energy modeling, David Greene's work represents both a response to Berry's fear that we still fail to appreciate the lasting impacts of the energy crisis, and Hogan's plea for better models to explain the impact of decisions. Greene's efforts to justify the need for fuel-efficiency standards stem from his examination of the impacts of OPEC on American society. Greene insists that unbeknownst to most energy market analysts, the 1973–74 price shocks, coupled with continued U.S. dependency on Middle Eastern oil, continue to pose a major public policy challenge.

The lesson of the price shocks is that energy markets alone may not be able to solve the problem of oil dependence or to produce a cost-effective level of energy efficiency.

In one sense, Greene's analysis brings the first two topics of this volume full circle. His conclusion more stridently acknowledges what Bohi and Darmstadter's historical retrospective only grudgingly concedes: that some programs born during the aftermath of the energy crisis have had both a positive and a durable impact. Improvements to energy efficiency and greater free-market competition have both enhanced energy efficiency. The energy crisis shows us that government intervention and free markets are both necessary to provide us with a stable, secure, and environmentally appropriate energy supply.

Unresolved Issues: Energy and Environment

Society's need to balance energy consumption and environmental protection requires a role, however unclear, for government. Over the next twenty years, and likely beyond, the battle over energy supply and environmental protection will continue to rage in developing nations. The World Bank's Dennis Anderson (chap. 11) ponders the chances that less developed nations can improve both their economic development through energy advancements and their environmental quality. While the litany of problems faced by these countries is overwhelming, compounding matters is their claim that they are at the mercy of stronger, developed nations for new sustainable development technologies. It is in the interest of developed nations to provide this assistance, because all nations will share the environmental benefits. The upshot of Anderson's argument is that such assistance is not a cost, but an investment.

For example, Americans are responsible for nearly one-quarter of all greenhouse gas emissions, most of which stem from the burning of fossil fuels. However, developing nations are catching up rapidly, having doubled their energy consumption between 1970 and 1990. The task of reducing consumption constitutes a more daunting challenge than we have been led to believe, because contrary to conventional wisdom, poorer countries may be more inclined, owing to a lack of technology, to consume biofuels such as dung, crop residues, and fuel wood in environmentally inappropriate or unhealthful ways (e.g., open burning, clear-cutting). Therefore, developed nations owe it to themselves to provide less developed countries with energy technologies that use biofuels in more environmentally benign ways. It is also necessary to promote lower fertility rates by improving infant and maternal health care, and to promote lower infant mortality in order to discourage large families.

Former Environmental Protection Agency official Richard Morgenstern, while sympathetic to all these ideas, suggests that we need innovative alternatives, including having all countries incorporate into the gross domestic product (GDP) indices measures of "green performance," to better aid decision makers in making sustainable policies. Governments could also suggest standards for desirable energy performance while permitting businesses to choose their own means to meet them.

Finally, the University of Tennessee's Robert Bohm suggests that we should not ignore the benefits of technological eclecticism. The end of the Cold War and the decline of superpower rivalries mean that collaboration and cooperation not thought possible twenty years ago can now be exploited. While countries' views on energy equity will continue to diverge, avenues for working together in an environmentally responsible manner have never been clearer, as the 1992 Earth Summit and its aftermath (see part 4) show.

The Future: "Déjà-vu All over Again"

What do the *next* twenty years portend for political stability, environmental quality, and economic justice if we continue to pursue current energy policies? And given changes since 1973–74, could such a crisis recur, and could the United States effectively recover from it? Our authors agree on four issues.

First, it is unlikely that such a crisis can recur, because of increased flexibility and diversity of supply and more abundant energy reserves. Long before a shortage would occur, energy prices would signal an adjustment by rising. Second, should another embargo occur, our ability to sustain the shock to our economy remains in doubt because energy prices are likely to rise, and gross national product (GNP) and average wages are likely to decline. Worse, the magnitude of these impacts remains uncertain. Third, improvements to energy efficiency and demand-side management inhibit (but do not prevent) such a crisis from recurring. It is simply not clear that a highly tuned energy-efficient economy is immune from supply disruption. Finally, assuming that this prognosis is correct, we should be adopting a "no-regrets" energy policy that provides diversification of supply and flexibility in managing demand. While a Strategic Petroleum Reserve able to serve as a backstop in case of foreign supply interruption can be a positive form of government intervention, such a reserve is by itself insufficient.

The contributors to this volume endorse a sustainable energy supply strategy. To the Electric Power Research Institute's Chauncey Starr (chap. 14), such an energy supply strategy would consist of technologies that can be rapidly transferred and introduced into countries that lack the money to develop alternative sources of energy on their own. In the short and medium term, such energy technologies might include nuclear power as well as fossil fuel plants. In the longer term, however, Starr and most of the other contributors would prefer to see other technologies developed whose impacts are environmentally more benign and that are cheaper in cost.

Flavin contends that a more desirable alternative to nuclear power would be identifying opportunities to use fossil fuels more innovatively. Hopeful options include the use of hydrogen fuel, natural gas, and hythane (a combination of methane and hydrogen that could be used to generate electricity and to heat homes). Electrolysis could be used to separate hydrogen from water, according to Flavin, while photovoltaic and other renewable energy sources could supply the electricity for separating hydrogen from oxygen in water molecules. Flavin's optimism is based on the premise that because these fuels can be readily introduced into the existing energy transportation and distribution infrastructure, they would pose little disruption to established markets and could be readily adapted by developing as well as industrialized countries.

Kathryn Jackson and Verrill Norwood of the Tennessee Valley Authority (chap. 16) shed some light on this issue of infrastructure in their cogent review of the federal utility's efforts to introduce demand-side management and integrated resource-planning strategies into its service area. Jackson and Norwood note that "infrastructure" is itself a variable, not a constant, because utilities may commit themselves to major shifts in how they provide electricity.

Jackson and Norwood argue that a significant change is taking place at TVA, as in other energy production organizations both in the United States and elsewhere. Prior mandates to simply sell power are slowly being supplemented by greater emphasis on conservation, energy efficiency, and supply innovation. In short, like other energy actors, TVA is seeking to make adjustments to energy supply and to become more attuned to consumer demand through emphasizing integrated resource planning and other major changes in traditional business operations.

Policy Recommendations

Five major conclusions emerge from these discussions. First, the United States is still highly dependent upon foreign oil, importing more than half of all the petroleum it consumes. This percentage is nearly equal to that in 1979 and nearly 10 percent greater than that in 1973 (Carroll, 1991: 39; Fisher, 1992: 5). Moreover, as Greene insists, this dependency continues to pose a major public policy challenge because it may result in a loss of potential

GNP and, thus, a loss of ability to produce. It also leads to a transfer of wealth to oil-producing nations. In essence, as a major consuming nation, we remain vulnerable to supply disruption.

Second, while energy prices have been deregulated and progress has been made in diversifying energy sources, the United States may still remain vulnerable to the conditions that led to the 1973 shock. While the authors represented here clearly disagree over whether OPEC really was an effective cartel, or whether rising prices for energy alone constituted incentives for conservation (see, e.g., Bohi and Darmstadter), they generally agree that our understanding of the complexities of international oil markets and the factors driving demand, supply, and decision-maker response leave much to be desired. For example, there is sharp disagreement over the necessity for, and usefulness of, stockpiling petroleum (e.g., the U.S. Strategic Petroleum Reserve) as a buffer against future cutoffs. As Canes notes, the motives that drive policymakers to impose "import quotas" or other ill-conceived policies intended to ensure development of domestic energy sources remain open to question.

Third, the authors agree that the United States should continue to diversify its energy sources in ways that make economic sense, such as moving toward non-fossil-fuel sources where appropriate and incorporating recent gains in the electronics and materials-science revolutions (particularly in transportation). These gains include using lighter-weight, energy-saving ceramics in engines, drivetrains, and car bodies, and more efficient "clean-burn" engine technologies. As Starr notes, there is a growing recognition that dependence on fossil fuels will not decrease in the near future.

The authors do not come to any agreement, however, as to what energy options are most appropriate for the developing world. While some, like Flavin and Anderson, contend that environmental and economic salvation lies in near-term development of clean coal and natural gas, as well as longer-term reliance upon community-based, decentralized renewable energy systems, others (e.g., Starr) suggest that it is premature to abandon central-station nuclear power plants as economical and practical alternatives, especially in light of developing countries' needs.

Fourth, the authors concur that greater effort must be devoted to understanding the complex effects of price, government subsidy, and public perception on energy R&D. Energy prices are not very good predictors of technological change or innovation. In short, there remains considerable debate on whether technology is exogenous (affected primarily by outside factors such as scientific inventiveness, public education, and public confidence with regard to change—e.g., Flavin) or endogenous (a product of changing market conditions dependent on price—e.g., Hogan). Moreover, there is strong resolve among divergent points of view to resist having government choose between "winners" and "losers." There appears to be a tension in government R&D policy between two goals. The first goal is to have an independent entity investigate the feasibility and absorb some of the financial risks of new technologies that are not yet ready for markets but whose development serves a public interest. The second goal is to ensure that government does not by itself take on the responsibility of making a new energy technology work. Most participants concur that government intervention is not an effective substitute for price incentives that make energy-saving technologies economically attractive and politically feasible. However, two contributors to this volume (Flavin and Greene) suggest that while this alternative is an ideal, some intervention is necessary to address market imperfections.

Finally, the authors agree that Americans have failed to wring every environmental and economic advantage from energy conservation. Despite considerable progress made by industry in improving the end-use efficiency of household appliances, automobiles, electric motors, lighting, and manufacturing processes, further improvements can be made, but only if energy R&D is guided by cost-effective, market-disciplined efforts, and not by "pork-barrel" politics.

The authors agree that the best course for energy policy is one that encourages technological developments that expand the range of plausible energy options. However, plausibility has its limits. None of these contributors points to nuclear fusion as a viable option looming on the horizon. It is simply too far removed from commercial viability.

Can energy R&D fulfill its promises? Despite continued skepticism, the answer is a resounding yes. Energy R&D was a significant component of federal responses to the two oil shocks of the 1970s, yet as Ellerman notes, it is given relatively little attention. This is due to the failure of R&D efforts by agencies such as the Energy Research and Development Administration that have emphasized unviable (e.g., synfuels) or unpopular (e.g., nuclear) power-generating technologies.

Between 1980 and 1990, energy R&D spending by DOE decreased by some 60 percent. According to one estimate, spending on solar and renewable energy during this same period declined by 93 percent (in constant dollars), while DOE's energy conservation budget fell by 91 percent (again, in constant dollars) between 1981 and 1987 (Kraft, 1991: 34). R&D improvements depend on two factors.

First, the feasibility of any new device, method, or source of energy will depend, ultimately, on the price of energy. Second, energy forecasting models are still not very good at encompassing rates of technological change and market penetration. In part this is due to the limitations of models. However, it is also due in part to the fact that many of the most relevant energy-use changes in technology are those that have occurred in such esoteric fields as electronics and materials science, whose rates of maturity, market penetration, and usability, as Flavin notes, are poorly understood.

Conclusions

More than twenty years have passed since the first energy shock—an interruption of supply precipitated only in part, we now realize, by deliberate foreign embargo. Changes in society have been profound. Although oil is actually cheaper today (in constant dollars) than in 1974, the transition in price precipitated in 1973–74 played havoc with the U.S. economy. The legacy of that havoc remains.

Motor vehicles are more fuel-efficient than in 1973. Gas and oil prices are largely deregulated, and there is far greater global diversity in the spot-market availability of oil. While the nuclear option appears moribund in this country, we still derive a large percentage of our power from light-water reactors. Moreover, in France and Japan, although orders for new plants have declined, a large proportion of electric power is still derived from light-water reactors. In France, a bastion of state support for nuclear power, political concerns over plant safety and waste management have arisen (Stetson, 1991; Ansel, 1992). Over time these concerns may have the same effect on nuclear power growth in that country that it has had in ours.

In the United States, there is now a Department of Energy, which is charged with, among other things, implementing a National Energy Strategy—a phrase that first became part of the national lexicon in 1974. And perhaps most significantly, as the Bush administration's cautious use of the Strategic Petroleum Reserve during the 1991 Persian Gulf War showed, policymakers at least are inclined to avoid capricious decisions (e.g., rationing, price controls) whose impacts may be worse than the problems they are designed to address.

More recently, the Clinton administration has proposed a Climate

Change Action Plan, one of the principal components of which is improvement in the energy efficiency of public buildings and transportation infrastructure (Clinton and Gore, 1993). These demands for improved energy efficiency were prompted by the perceived threat of global warming, an issue that was hardly on the environmental policy agenda in 1973. This change reflects the fact that policy concerns do respond to new scientific evidence.

Finally, and more soberingly, many of the vulnerabilities brought to the surface by the oil embargo remain with us. While OPEC is weaker than it was in 1973, the strategic interests of the United States are still defined in part by oil. Moreover, the appropriate societal response to this strategic reality, as Horwich, Bohi and Darmstadter, and Anderson note, still revolves partly around competing philosophies of the relative efficacy of market responses as opposed to government regulation. In short, despite changes in energy supply flexibility, international political conditions, environmental concerns, and energy efficiency, changes in energy demand have been less pronounced. We still rely heavily upon fossil fuels—and probably will continue to do so for the foreseeable future. And we still import a sizable portion of our oil.

Events since 1973 have taught us that government intervention in energy, as in other policy areas, is neither a cure-all nor a curse. In a very real sense, the energy debate that began in 1973 was a precursor to other debates over the proper role and scope of government in addressing social problems. For example, the current debate over health care in the United States revolves, at least in part, around precisely this issue. Should government's role be to control prices, ration supply, and ensure equitable distribution of health-care services by commanding their allocation? Or should government act to encourage economically efficient private-sector efforts such as "managed" health-care consortia composed of service providers and consumers? Federally imposed mandates to provide elderly and indigent care, without regard to rationing or price, are partly responsible for skyrocketing costs.

As Fox-Penner notes, the final legacy and lesson of the energy crisis may be our failure as a society to decide what's wrong with energy policy. Complex relationships between government intervention and markets suggest the need for "cautious activism," the hallmark, according to Fox-Penner, of the Clinton administration's energy policy.

But what would a cautious-activist policy look like? In short, where might energy policy be headed? It would be presumptuous to suggest that a wide-ranging volume representing diverse and sometimes divergent points of view could provide a simple answer to such a question. Nevertheless, some remarkably consistent themes emerge from this collection. One such theme is the importance of continued R&D and technology developments—the importance of endorsing a flexible energy policy that makes possible the capacity to produce things needed to confront future adversity. This includes the ability to encourage conservation, efficiency, and supply initiatives that have the chance of leaving options open to future generations that are at least no worse than we had left for ourselves.

NOTES

1. For additional detail on these activities, see chaps. 3, 4, and 9 in this volume.

2. Wellhead price controls were removed by Carter. Pipeline decontrols, removed by a combination of FERC (Federal Energy Regulatory Commission) and Reagan administration actions, permitted gas to flow virtually anywhere demand was manifested.

REFERENCES

Aman, Alfred C., Jr. 1983. *Energy and Natural Resources Law: The Regulatory Dialogue.* New York: Matthew Bender.

Ansel, P. 1992. "French Attitudes to Nuclear Energy in 1991." *Nuclear Europe Worldscan* 1–2:66.

Bohi, Douglas R., and Milton Russell. 1978. *Limiting Oil Imports: An Economic History and Analysis.* Baltimore: Johns Hopkins University Press, for Resources for the Future.

Carroll, John E. 1991. "Environmental Restraint: Living Lightly or Not Living." Pp. 37–46 in *Energy, the Environment, and Public Policy: Issues for the 1990s,* edited by David L. McKee. New York: Praeger.

Clinton, William J., and Albert Gore, Jr. 1993. *The Climate Change Action Plan.* Washington, D.C.: The White House.

Fisher, William L. 1992. "Energy Policy after Gulf War." *Forum for Applied Research and Public Policy* 7 (1): 5–11.

Kraft, Michael E. 1991. "Environmental and Energy Policy in the Reagan Administration: Implications for the 1990s." Pp. 19–36 in *Energy, the Environment, and Public Policy: Issues for the 1990s,* edited by David L. McKee. New York: Praeger.

Krasner, Stephen D. 1985. *Structural Conflict: The Third World against Global Liberalism.* Berkeley: University of California Press.

MacDonald, Gordon J. 1992. "Technology Transfer: The Climate Change Challenge." *Journal of Environment and Development* 1:1–40.

National Renewable Energy Laboratory, Solar Energy Research Institute. 1990. *The Potential of Renewable Energy: An Interlaboratory White Paper.* Golden, Colo.: National Renewable Energy Laboratory, March.

Nitze, W. 1990. *The Greenhouse Effect: Formulating a Convention.* London: Royal Institute of International Affairs.

Starling, Grover. 1979. *The Politics and Economics of Public Policy: An Introductory Analysis with Cases.* Homewood, Ill.: Dorsey Press.

Stetson, Marnie. 1991. "France's Tottering Nuclear Giant." *Worldwatch* 4 (1): 38.

Uslaner, Eric. 1989. *Shale Barrel Politics: Energy and Legislative Leadership.* Stanford, Calif.: Stanford University Press.

PART ONE

BACK TO THE FUTURE
Causes and Consequences of the 1973–1974 Energy Crisis

Did government intervention cause, help to resolve, or merely aggravate the impacts of the 1973–74 Arab oil embargo? On the surface, the five chapters in this part argue for the latter: U.S. government policies worsened the impacts of the energy crisis through oil and gas price and allocation controls. Ironically, oil price controls were introduced by the Nixon administration ostensibly to relieve inflationary pressures by channeling demand toward imports. However, the authors of these chapters also contend that any final verdict on the role of government and private markets in the origins and resolution of the energy crisis must be more ambivalent than this superficial view would alone suggest.

Doug Bohi and Joel Darmstadter, of Resources for the Future, draw a compelling portrait of energy policy in the years immediately preceding the crisis. They find unmistakable evidence that government-imposed restrictions on energy markets exacerbated the Arab oil embargo. Particularly noteworthy was the entitlement program under which refiners with access to cheap domestically produced oil subsidized others that were dependent on more costly foreign oil. Government actions following the inception of the crisis were not much better thought out, Bohi and Darmstadter contend. For example, an ill-fated synthetic fuels program languished for want of an open market generating long-term demand and realistic price incentives.

However, Bohi and Darmstadter further suggest that other programs born during the aftermath of the energy crisis have had both a more positive and a more durable impact. Mandated vehicle fuel-efficiency standards, renewed competition in the electricity generation industry, and oil and gas price decontrols—all of which would probably not have come about without the energy crisis—have helped to ensure a more stable energy supply and greater institutionalization of energy efficiency. In short, the energy crisis must be partly credited with improving our understanding of the environmental and societal consequences of energy use. While many of the challenges posed by these interrelationships remain unresolved, at least there is an emerging consensus about the need to combine market incentives with government prodding.

Denny Ellerman, of the Massachusetts Institute of Technology, reinforces this conclusion by suggesting that the most enduring impact of the energy crisis is the loss of public confidence in government as a remedy for market deficiencies. Ultimately, contends Ellerman, this erosion of faith in government intervention is due to (1) the failure of many energy policies that had their genesis in the 1970s to deliver on their promises (e.g., to provide for energy independence and to make available economically competitive renewable energy technologies); and (2) the perception that these policies were promoted at the behest of private rather than public interest.

To the extent that Americans continue to view government as beholden to "special" interests that seek taxpayer subsidies or market protection, they will continue to regard government intervention in energy with warranted skepticism. Louisiana State University's Allan Pulsipher focuses upon the ideological reasons for government intervention in energy markets as a vehicle for explaining what he believes to be the failure of intervention. Employing the psychological metaphor of hallucination—a "distortion of, and loss of contact with, reality"—Pulsipher whimsically suggests that many government policies adopted since the oil embargo constitute both a departure from historic laissez-faire policies and a misplaced faith in so-called soft energy technologies.

It is not the technologies themselves that draw Pulsipher's wrath, but the inconsistencies in their proponents' endorsements of them. How can it be, Pulsipher rhetorically asks, that great progress has presumably been made in technologies that permit decentralized energy use and greater conservation, while larger, "harder" technologies (e.g., nuclear power) continue to languish despite considerable R&D efforts? The answer put forth by proponents, as justification for greater government intervention, seems to Pulsipher especially problematic. They contend that the public's gross ignorance about renewable technologies is nurtured by "disinformation" fostered by energy interests and government agencies. Ironically, these staunch advocates of renewable and alternative energy sources seem to share with free-market conservatives a distrust of government. Whereas the latter view the policy process as beholden to entrenched interests, the former see it as part of a deeper conspiracy to deny a market for new technologies and to keep the public literally in the dark.

Purdue University's George Horwich reminds us that if certain advocates of renewable energy technologies are guilty of pleading for more, not less, regulation, they are not alone. Before, during, and after the 1973–74 energy crisis, many economists and policy professionals supported price controls and mandatory allocations, arguing that they were necessary to ensure a stable, reliable energy supply. Beginning with President Jimmy Carter's Council of Economic Advisers, some policymakers have feared that in the absence of controls, an oil supply disruption could dramatically raise prices, unleash massive consumer spending on oil products, and create producer profit windfalls. Moreover, some view allocation controls as protection for smaller refiners, which can be subsidized with crude oil supplies during times of market interruption.

Carefully documenting the inefficiencies generated by such policies, Horwich notes that more effective market "stabilizers" are available in times of crisis. As Operation Desert Shield/Desert Storm showed, the effort to deny the benefits of control of Middle Eastern oil to Iraq's Saddam Hussein was a prudent, if ambivalently followed, course of action that relied on a combination of free markets and military intervention. Given that Saddam "would surely have restricted [oil] output below previous levels" had he remained in control of Kuwaiti oil, he was a palpable threat to the security of the Middle East and to ourselves. In short, one need not be an emissary for "new" energy sources to find an appropriate role for government intervention.

This retrospective concludes, appropriately, with the *Washington Post's* John Berry. While largely concurring with the foregoing analyses, Berry focuses on the political damage wrought by the embargo and its aftermath. The energy crisis contributed to a persistent rise in inflation that led to three recessions in the span of eight years. Aside from increasing this nation's balance-of-payments deficit, wreaking havoc on the banking and thrift industries, and exacerbating the national debt of less developed countries, the oil embargo's spillover effects led to the fall of one president (Jimmy Carter), influenced the fiscal and monetary policies of two others (Ronald Reagan and George Bush), and may have given birth to the catchword *policy gridlock*.

While the underlying causes of these problems remain open to contention, Berry makes a compelling case for the argument that had the oil embargo not occurred, and had the government been better prepared for the 1979 price shocks, subsequent history might have been much different. Underscoring a point made by several other contributors to this part, Berry concludes that not only do we ignore the lessons of the embargo at our peril but, on a more positive note, various recent actions—by the Federal Reserve to control energy-driven inflation in 1990–91; by the U.S. Department of Energy to rely upon greater market incentives to increase access to petroleum supplies; and by the Bush administration to avoid rash measures to control petroleum markets following Iraq's seizure of Kuwait—signify appreciation for the failure of previous policies and a greater regard for modest alternatives.

2. The Energy Upheavals of the 1970s

Policy Watershed or Aberration?

DOUGLAS R. BOHI AND JOEL DARMSTADTER

Were the oil price shocks of 1973–74 and 1979–80 "defining moments" in economic, social, and political affairs? Did this decade leave in its wake a new set of institutional imperatives, arrangements, and ways of thinking about the world? Or was it an incidental departure from longer-term trends? When the dust finally settled, new realities, concerns, and perceptions emerged. At the same time, streaks of continuity remained, proving that what is and is not new does not always conform to commonly held perceptions.

We have learned a great deal since the oil shocks. This becomes quickly apparent as one faces up to early misjudgments about OPEC's robustness, the failure to recognize our limited capacity to simulate energy market behavior, misplaced confidence in government's ability to influence energy policy along socially efficient, desirable paths, and our pervasive reluctance to face up to the enormous uncertainty shrouding these issues. The record is highly flawed. In reviewing this era it is sometimes difficult to retain the scholarly equivalent of a straight face.

Our instincts are to avoid labeling the events of the 1970s an energy "crisis," taking that characterization—analogous to its clinical usage—to mean a critical turning point in the condition under examination. But we use this phrase partly out of habit and convenience, and partly because it would be somewhat disingenuous not to. At the time, these perturbations were widely perceived as crises, and they decisively influenced public policy. As later research conducted in a less feverish atmosphere began to dissect the 1970s with more faithful attention to the economic and political forces under which they operated, the "crisis" interpretation of events lost credibility.

OPEC and Antecedents to Events of the 1970s

Many writers have held that the Arab oil embargo of 1973–74 and the post–
Iranian Revolution years 1979–80 occurred within an institutional framework defined by a shift in bargaining strength from Western oil companies
to largely Middle Eastern oil-producing governments. Years later, reinter-
pretation of these events assigned a greater role to markets. Even today a
totally satisfying explanation of how the oil price explosions of the 1970s
came about cannot entirely neglect the role of OPEC. In turn, the events of
the 1970s cannot be fully grasped without referring to developments over
the prior half-century.

For most of this period a key determinant of world oil market conditions
was the degree of control over supply, price, and market shares exercised by
the multinational corporations—the so-called Seven Sisters. In addition,
after World War II the U.S. government, intent on safeguarding access to
the rapidly growing and presumably indispensable oil reserves of the Middle East, worked out favorable tax treatment of royalties paid by conces-
sionary companies to governments. Incipient signs of the nationalism that
fully flowered with state takeover of foreign oil properties in the 1960s and
1970s were also evident soon after the war. For example, the 1950s saw the
nationalization of the Suez Canal and the Iranian oil industry.

Notwithstanding doubts about the extent of its power, the establishment
of OPEC in 1960 must be figured as a notable milestone. Weak world oil
prices significantly conditioned by U.S. import quotas adopted in 1959
provided the right background conditions. The immediate cause of OPEC's
formation was an attempt by the multinational oil producers unilaterally to
reduce the posted (or tax reference) price of crude oil by up to 8 percent.
While a comparatively modest decrease, it was viewed as a "last straw"
by Venezuela and several other major producers, who joined forces and
founded OPEC. As a significant oil exporter to the United States, Venezuela,
in addition, felt victimized by U.S. oil import quotas introduced in 1959 (see
Bohi and Russell, 1978).

Throughout the 1960s, OPEC's role expanded rhetorically as well as sub-
stantively. It voiced its intention to secure equity participation in the producing areas and to have a major say in determining posted prices. At the
same time, Western companies lost some of their bargaining cohesion,
partly because of the growing and discordant presence of independents
such as Occidental Petroleum. They submitted to rising price demands by
several OPEC members. Market conditions accommodated these develop-

ments. Throughout the decade oil consumption outpaced production. And so it was not merely the October 1973 Yom Kippur War but the broader market, political, and regulatory realities that had been building for some time that allowed OPEC to quadruple oil prices (from $3 to $12/barrel) within a three-month period in the winter of 1973–74.

The Arab oil producers' embargo, aimed principally at the United States, turned out to be somewhat anticlimactic—an act further reflecting the shifting realignment of the world oil scene. In terms of its effect, however, it was largely a political gesture. It failed to penalize individual targets of the embargo, nor did it materially aggravate the prevailing global oil supply situation. On the contrary, the embargo provided the motivation for greater exploration for new supplies around the world and a shift away from oil in all but the transportation sector.

The Prevailing Regulatory Environment

While the energy crisis of 1973–74 is commonly thought to have been caused by the Arab oil embargo, it was actually the result of many separate and unexpected events that simultaneously converged. In the United States, energy supply was constrained and growth in demand for imported oil was unexpectedly high. These developments were largely the product of energy policies pursued for many years prior to the 1970s. Some of these policies were implemented in the name of energy security.

Regulations imposed on the natural gas and coal industries reduced the availability of substitutes for oil and forced increases in energy demand to favor oil. The field price of natural gas had been regulated for two decades before the energy crisis. By holding the price below its market value and replacement cost, these regulations reduced the incentive for new discoveries, increased gas consumption, and distorted the allocation of supplies among consumers. The growing demand for gas caused the price to rise in unregulated intrastate markets, which absorbed new supplies, and forced consumers in regulated interstate markets to switch to oil products. This accelerated the demand for oil in the years leading up to the worldwide pressure on oil supplies in the 1970s.

The coal industry was also in poor condition to meet growing demand and to relieve the pressure on oil. That industry suffered from decades of decline exacerbated by low gas prices that left it unable to respond to the energy shortages of the 1970s. Just when the shortages of natural gas developed in the late 1960s, the nation began to make up for years of environmental neglect and poor health and safety conditions in the coal industry. Implementation of new environmental, health, and safety regulations initially raised the cost of coal production and lowered labor productivity. Thus, coal was not readily available as a substitute for oil and gas when needed.

The events in gas and coal focused energy demand in the direction of oil, and developments in the U.S. oil industry channeled that demand onto the world oil market. In the decade prior to the Arab oil embargo, the United States constrained oil imports with a quota program that limited their volume to approximately 12.2 percent of domestic production. When the quota was implemented in 1959, the world price of oil was almost \$1/barrel below the domestic price of about \$2.80/barrel. Imports comprised an increasing share of the domestic market for several years, and the point had been reached where President Dwight Eisenhower determined that the domestic petroleum industry—and thus national security—would be threatened by any further encroachment. The quota program stopped the growth of oil imports and stabilized the price of domestic oil until 1970, when it became increasingly evident that quotas could no longer be maintained. Restrictions were eased in 1971 and formally removed in April 1973, just months before the Arab oil embargo (cf. Bohi and Russell, 1978).

To understand why the quota program ultimately failed, one must understand the long history of oil price controls imposed first by state governments in the name of resource conservation and later by the federal government in the name of anti-inflation. Oil-producing states responded to increasing production and falling prices in the 1930s with a series of conservation measures that included what is known as *demand prorationing*.

An estimate of the quantity demanded at a prescribed price level determined maximum permitted oil production. Total production was then rationed among all wells by imposing a production limit on each. This system ensured that excess production would not occur and that the domestic price of oil would not fall. The federal government ratified this system of price controls by prohibiting the interstate sale of oil produced in violation of state regulation.

This system worked smoothly until the 1950s, when major oil discoveries in Africa and the Middle East undermined the world price of oil and increased incentives for importing foreign oil into the United States. As oil imports grew, the share of domestic demand supplied by domestic sources shrank. Political pressure for oil import controls grew until the quota system was implemented in 1959. The quota program thus sustained the stable level of oil prices that had existed in the 1930s despite subsequently steadily growing demand. Production controls, however, undermined the incentive for investing in new reserves to replace those being rapidly depleted. In the meantime, consumers responded to decades of stable prices—albeit at levels above uncontrolled market conditions—as if supplies were unlimited and available at constant cost.

The appearance of stability began to unravel, not because of a worldwide energy crisis, but because of domestic inflation. Inflation became a matter of national concern in 1968 and 1969. In 1969 the government responded with a program of monetary and fiscal restraint that resulted in a decline in the growth of GDP and a rise in unemployment in 1970–71. Nevertheless, inflation persisted and the government began to explain the phenomenon of "stagflation" in terms of inflationary expectations: past experience with rising prices leads businesses and workers to expect the trend to continue. Thus, they behave in ways that realize their expectations. Government controls on wages and prices are needed to break the expectations spiral.

Employing this reasoning, the government began a ninety-day freeze in August 1971, followed by regulations that lasted until January 1973 limiting annual price increases to 2–3 percent. It was during this period that shortages of petroleum products began to occur and increases in imports were allowed. In response to growing worldwide demand, the price of oil also increased. Under these conditions the ultimate unworkability of the price control program became evident. For example, rising import costs could not be passed on to consumers unless foreign products were kept physically separate from domestic ones. The supply system was incapable of assuring the separation, so imports of key products in short supply were discouraged. Allocation controls, in the meantime, restricted the supply of no. 2 fuel oil (used mostly for home heating) more stringently than the supply of diesel fuel. As a consequence, since the two are chemically equal, some fuel oil consumers began to burn diesel fuel despite its higher price and added highway taxes.

After January 1973, price controls became self-administered by firms and were ended altogether when the Economic Stabilization Act expired in April 1974. Regulation of oil product prices continued. By 1974, however, world oil prices exceeded domestic prices, so it became impossible to import at controlled domestic prices. Imported products had to be exempted from the price controls so that they could relieve the shortages in domestic energy supplies. This was accomplished with what became known as the "entitlements program," which in the course of a year transformed the

United States from a country that officially restricted oil imports to one that subsidized them (and which, coincidentally, increased world oil demand).

The entitlements program was intended to equalize the cost of oil acquired by domestic refiners, and it worked like a mirror image of the earlier oil import quota program. Whereas the quota program had restricted imports to a percentage of domestic production, the entitlements program allowed importing refiners to receive a subsidy in relation to the share of costlier oil imports they refined. Refiners who exceeded the national average of price-controlled domestic oil had to pay refiners who produced below the national average of domestic oil. Per-barrel payment was based on the difference between domestic and foreign oil prices.

These crude oil pricing and entitlement provisions were superimposed over already tortuous cost pass-through and allocation requirements on the petroleum industry. By late 1975 there was widespread uncertainty about production costs and legal prices, and about whether any company's operations complied with regulations as they might be interpreted. The end of the quota program was preceded and followed by pervasive government intervention in petroleum pricing and allocation.

Formally, the end of oil import quotas in 1973 was marked by their replacement with import fees. It became evident that the nation was unable to meet its energy needs in the face of both oil price controls—which remained in place for the rest of the decade—and import quotas: "incentives" reduced domestic energy production and increased domestic energy consumption, while controls limited the ability of imports to close the gap. The new import fee system that replaced quotas did not represent a new departure in the taxation of oil imports. On the contrary, oil import fees had been in place since 1932, and the new system merely extended existing fees. Ironically, the new fee structure lowered the total levy on imports. Thus, in 1973 oil imports returned to the formally uncontrolled status that had existed before 1959.

Pre-1973 Environmental Stirrings

Without minimizing the importance of what happened in the 1970s, we must acknowledge that stirrings about environmental threats, resource scarcity, and other energy-related problems clearly predated the 1973 oil shock. A few examples attest to how visibly things had churned on both the policy and intellectual fronts (Greenberger, 1983). The decade preceding the 1973–74 oil price upheaval saw the enactment of the following:

—the original (1963) and amended (1970) Clean Air Acts, providing for air pollution abatement
—the Mine Safety and Health Act (1969), which, despite weak enforcement for several years, established federal authority for promulgating and enforcing miner health and safety standards
—the National Environmental Policy Act (1969), creating a Council on Environmental Quality and requiring environmental-impact statements (which a federal court ruled in the Calvert Cliffs case applied to nuclear power plants)

More squarely centered on energy, and reflecting anxiety over short- and long-term supply, were the Trans-Alaska Pipeline (TAP) Authority Act (1973), facilitating construction and limiting legal challenges to TAP and the export of North Slope oil, and the pre-embargo Emergency Petroleum Allocation Act of 1973, which controlled price and allocation.

Intellectual and ideological ferment during this period was highlighted by the publication of *The Limits to Growth* (Meadows et al., 1972). The medium for this study was a computerized, dynamic model of the global environment whose principal variables included demographics, agricultural output, nonrenewable energy and other resources, industrial production, and emissions. Its message was that trends pointed to an "overshoot and collapse" scenario over the next century or so. Avoidance of collapse could not be ensured by traditional recourse to advancing technology and substitution of abundant for scarce resources, but required "entirely new approaches to redirect society toward goals of equilibrium rather than growth" (Commentary by the Club of Rome Executive Committee, in Meadows et al., 1972: 197).

Only a slightly less frightening vision of the future came from Barry Commoner, who, at a 1971 Resources for the Future (RFF) forum, declared that "if we are to survive economically as well as biologically, much of the technological transformation of the U.S. economy since 1946 will need to be . . . redone in order to bring the nation's productive technology much more closely into harmony with the inescapable demands of the ecosystem" (Commoner, 1972: 65).[1]

While these authors veered toward an overly alarmist depiction of the human prospect, serious concern was also expressed in more dispassionate analyses. Two studies sponsored by the Massachusetts Institute of Technology—*Man's Impact on the Global Environment* (MIT, 1970) and *Inadvertent Climate Modification* (MIT, 1971)—are good examples. Energy and its environmental consequences figured prominently in both. The first of these reports concluded "that [while] the probability of direct climate change *in this century* resulting from CO_2 is small, we stress that the long-term potential consequences of CO_2 effects on the climate or of societal reaction to such threats are so serious that much more must be learned about future trends of climate change. Only through these measures can societies hope to have time to adjust to changes that may ultimately be necessary" (MIT, 1970: 12).

In short, scholarship in the years preceding the 1973–74 energy disruption reflected growing concerns over the adequacy, use, and external impacts of energy and other resources. The postembargo period saw a dramatic upsurge in these concerns.

Post–Oil Shock Approaches

The 1973–74 oil price shock set in motion debate, policy analysis, and theoretical studies on energy topics. These efforts gathered additional momentum in the wake of the 1979 Iranian Revolution, when oil prices again increased sharply—doubling to approximately \$24/barrel.[2] This frenetic spurt of activity was driven by a presumption by many prominent analysts that substantial and steadily rising real oil prices were inevitable; by the belief that arresting this and other countries' rapidly rising energy needs would threaten major economic harm; and by the conviction that oil consumers throughout the world were in the grip of a cartel with a demonstrated capacity and will to wreak havoc upon the international economy. Other analysts went further. Notwithstanding vast differences between oil and other resources with respect to substitutability and geographic dispersion—and notwithstanding the futility of past cartelization efforts—some speculated that OPEC was a harbinger of commodity cartels, a prospect justifying defensive strategies by consumer nations.

A prominent voice along these lines was Fred Bergsten, then of the Brookings Institution, who stated flatly before a congressional committee not long after the embargo, "There can now be no doubt that a large number of primary producing countries will be making steady, determined and often concerted efforts to raise substantially their returns from a wide range of commodities which they produce, including through the formation of new OPECs and . . . that many of them are in an excellent position to do so" (U.S. Congress, House of Representatives, 1974: 112). An opposite view was expressed by Hans Landsberg of RFF, who, after analyzing significant differences between OPEC and other primary producers, testified, "I do not think it altogether useful to conjure up the idea of 'more OPECs.' I do think it useful to discuss possibilities of attempts by exporters to manage raw materials in order to get substantially higher prices. But the analogy to OPEC is a bad one and is one I would not like to have as a popular image. It sends us off in a direction in which the parallels simply are dim or do not exist" (p. 133).

In the unsettled mood of 1974, the notion of forestalling possible supply and price manipulation by primary commodity exporters—while acknowledging the perennial problem of price volatility experienced by these exporters, especially in the developing world—was echoed by Secretary of State Henry Kissinger. In what appeared to be a reversal of this country's general aversion to international commodity price agreements, the secretary indicated in appearances both before the Organization for Economic Cooperation and Development (OECD) and the United Nations General Assembly in 1975 a willingness to consider them on a case-by-case basis (see testimony by various witnesses, U.S. Congress, House of Representatives, 1976).

Few contributions to energy analysis and debate in the 1970s were immune to the powerful temptation to project perceived trends and overreact to prevailing circumstances and anxiety. And so the somewhat frenzied mood of the 1970s and early 1980s produced the demand that "we do something—anything!" to head off these scary scenarios. Moreover, the perceived need for change centered not merely on oil but on energy in its broadest dimensions: the mix of sources, security questions, resource scarcity, technological progress, lifestyles, environment, economic growth and welfare, and, cutting across all these issues, the nontrivial matter of defining the appropriate role of government in ensuring the best possible outcomes. The major responses to the energy crises of the 1970s may be collapsed into five broad headings, as follows.

Import Dependence and Energy Security

While the immediate problem facing the United States after the 1973–74 Arab oil embargo was how to increase oil supplies to relieve energy shortages created by price and allocation controls, the longer-term problem was energy security. The Nixon administration responded to this problem in two ways: an international plan, led by Secretary of State Kissinger, aimed at countering the market power of OPEC, and a domestic plan, led by Federal Energy Office head William Simon, aimed at eliminating dependence on foreign oil.

Kissinger's plan assumed that the high price of oil was due to OPEC production restrictions. Consuming nations had to form a unified block in order to confront OPEC at the bargaining table. Negotiation was seen as the way to stabilize the world oil market, regardless of the underlying economic conditions that led to the crisis. Twenty-two members of the OECD (France and Iceland declined) formed the International Energy Agreement, which created the International Energy Agency (IEA) in 1974. The principal component of the agreement was a plan to reduce oil import demand at a time of world market shortage through conservation, import control, or stock drawdown. Because most other countries in the group lacked domestic production capacity, and therefore were not in the same bargaining position as the United States, an important element in the unification effort was a guarantee to share American oil in the event of an OPEC-engineered supply reduction.

Although the IEA still exists, its influence in international oil market conditions and developments has been minimal. To some extent this impotence is the result of the incorrect assumptions that led to its formation. As the passage of time demonstrated, OPEC has not had complete control of oil prices, and oil producers were as much subject to market conditions as oil consumers. Indeed, OPEC proved to be far from cohesive. Members never reached agreement on production quotas. This was especially true during the critical period between 1981 and 1986 when demand and the market price continually fell.

At the same time, oil-consuming countries also proved to be less than unified. The United States insisted that IEA members build strategic oil stocks in order to meet their commitment to deal with supply disruptions. Other countries insisted that they could not afford stockbuilding efforts sufficient to make a difference and pledged to reduce consumption when the need arose. The measures these countries proposed to use to reduce consumption did not invoke confidence in their commitment, however. They were based on ineffective, indirect approaches and moral suasion rather than taxes and quantity restrictions. While the United States embarked on a $20 billion program to build an oil stockpile, other countries have made little commitment to do the same.

A similar problem of commitment arose in U.S. policy, reflected in the inconsistency between Kissinger's strategy of sharing oil with other importing countries and President Richard Nixon's goal of energy independence. The independence goal suggested that the United States might go it alone rather than try to hold the consuming block together. Also, energy independence would have a chance of succeeding only in a world of high energy prices, while the Kissinger plan was aimed in the opposite direction.

Although it turned out to be just one in a succession of policy initiatives introduced over the course of the next few years, President Nixon's Project Independence, introduced in January 1974, was the centerpiece of domestic energy policy. From the start it was clear that high energy prices would be necessary to encourage conservation and domestic production so as to achieve energy self-sufficiency. As the analysis of the objective proceeded, it became apparent that the necessary increase in prices would be too large. Complete self-sufficiency was so costly that it was soon regarded as unfeasible. Consequently, the goal shifted from independence to "limited vulnerability." An essential part of this goal was reducing imports. Thus, oil imports were viewed as the measure of energy insecurity, and their reduction was, and continues to be, the underlying objective of energy policy.

Altering the Energy Mix

Different energy sources are substitutable over a wide range. The extent to which one form of energy actually replaces another depends on relative price, technology, location, government policy, and other factors. However, the persistence of interfuel competition in, for example, electric generation and space heating attests to the importance of substitutability. The oil price escalation of the 1970s and the concern with undue oil import reliance directed the attention of analysts and policymakers to the question of how to alter the energy mix significantly away from petroleum. Particular emphasis centered on nuclear power, coal, and oil shale. A major 1976 report by the Federal Energy Administration (a predecessor agency to the U.S. Department of Energy) stated, "Since the nation has taken a strong position in favor of reducing our reliance on imported oil, one of the major alternatives available is to develop a new industry that will make synthetic fuels (oil and gas) from abundant indigenous resources of coal and oil shale" (U.S. FEA, 1976: 315).

Such bullishness, predicated on the prospect of synfuel costs being within a range likely to be approached and perhaps surpassed by world oil prices in the not-too-distant future, was not limited to governmental pronouncements and analyses. A 1979 RFF study suggested that oil shale and other synthetics "appear to promise ample substitutes, at real costs about double 1978 prices of imported oil" (Schurr et al., 1979: 56). This would translate into shale oil available in the $25–30/barrel price range (or in the $40–50/barrel range in 1992 prices) when oil prices hovered around $20/barrel, unchanged in real terms from their late 1970s level prior to the Iranian Revolution (RFF envisioned a doubled real oil price materializing soon after 2000).

In the interim, efforts by the federal government to promote, and by private industry to establish, a significant shale-oil-producing capacity evaporated. In 1981 Exxon committed nearly $2 billion to its Parachute, Colorado, oil shale facility "to produce 50,000 barrels of oil a day from . . . oil shale, with production start-up targeted for late 1985" (Exxon Corporation, 1981). Within a couple of years the company had spent, according to one report, more than $900 million in Colorado alone (Kirkland, 1984: 31). Other estimates put the amount substantially lower.

In any case, by early 1982 Exxon had shelved this and most of its other synfuels projects as production costs appeared to be no less than $60/barrel, or close to $80/barrel in 1992 prices. Success also eluded other companies (e.g., Unocal), since neither government price nor loan guarantees were sufficient to overcome high production costs. But then, throughout much of the twentieth century, shale oil has perpetually been "just out of reach" of the competitive zone defined by crude oil.[3]

During this period the gap widened sensationally in spite of the federal government's best effort—through financial support under the Energy Security Act of 1980, creating the now-defunct Synthetic Fuels Corporation— to spur a synfuels production capacity of several million barrels per day crude oil equivalent by 1992. In Exxon's case, no federal funds were involved; even hard-nosed private entrepreneurial strategies can flounder, though not often with such a precipitous about-face.

The Emergence of Environmental Concerns

Environmental protection was viewed as a serious social issue prior to the first oil crisis. But the energy disruptions of the 1970s and the debate over proposed means of avoiding a recurrence helped bring environmental concerns even more emphatically into the spotlight. Each major petroleum alternative was seen as incriminated on environmental grounds:

—Coal was unsafe to mine, left polluting spoils, and, when burned, released harmful combustion products.

—Synfuels disturbed the land, contaminated water, and—in the context of climate-change concerns beginning to be addressed—were a potential source of greenhouse gas emissions much greater than the conventional fossil fuels they would replace.

—Nuclear power, chronically confronted by critics over waste management and proliferation dilemmas, found itself having to contend with the implications of the 1979 Three Mile Island accident. The plutonium-generating breeder reactor, whose governmental R&D support effectively ended in 1983, was perceived as representing an even more dangerous long-term course, even though it promised a virtually infinite fuel supply.

The challenge of how to deal with these and other environmental fears should, of course, not be framed too rigidly. *Some* energy-connected environmental harm was inescapable, irrespective of the energy system. On the other hand, at some cost (e.g., by switching from high- to low-sulfur coal) the benefits of energy use could be realized with muted environmental impact. By the 1970s economists were addressing some of the paradoxes of industrial capitalism in which markets work efficiently in mediating most transactions but poorly in forcing polluters to bear the cost of their assault on common-property resources and other assets external to their production activity. Taxes or other measures to attenuate environmentally damaging behavior existed. Note Kenneth Boulding's 1971 poem, which included the couplet "Economists argue that all the world lacks is / A suitable system of effluent taxes" (Boulding, 1972: 139).

In spite of such enlightened approaches to managing environmental problems, suspicions lurked that these approaches would never be in the vanguard of energy policy but would always remain a marginal force. The nuclear bandwagon might be unstoppable. Expanded coal production and promotion of a synfuels industry could easily override concerns about associated social costs. In any case, environmental concerns over these and other sources, none of which lacked undesirable side effects, would grow apace. Hence, it appeared to many that a more basic technical, economic, and sociological question with strong normative overtones must be posed: How much energy was really required to accommodate the lifestyle, welfare, and material aspirations of society?

Conservation, Lifestyles, and Economic Growth

The 1970s gave rise to broad-based reflections on the links between desirable socioeconomic goals and energy requirements. The traditional and misguided paradigm, those arguing for a fundamentally new world view asserted, proceeded something like this: Rapidly rising levels of per capita national product—too often mistaken for advances in human welfare—involve virtually lockstep growth in affordable energy supplies. This is because the nation's factories, farms, and other production activities depend on fuels and power to sustain and expand output, and because rising income produced by rising output is partly spent on creature comforts (e.g., cars and appliances). Critics felt that both propositions, promoted in large measure by those pushing to enhance energy supply as opposed to the more muted voices emphasizing frugality and conservation, were substantially flawed. New paradigms were urgently needed. If the marketplace was ineffective in stimulating restraint in energy use, it was thought that public policy could coax things in the right direction.

The stylized and widely promulgated depiction of the lockstep energy/GDP relationship was seriously distorted. Long before the highly vocal exchanges on that question in the 1970s and 1980s, an RFF study showed the influence of technological factors in bringing about changes in the relationship between energy and national output—an influence frequently powerful enough to mask the putative effect of price. Thus, for several decades following World War I, U.S. energy consumption trailed GDP growth even as *falling* real energy prices might have been expected to encourage the opposite result (Schurr and Netschert, 1960). The spread of energy-related technologies (e.g., factory electrification, farm mechanization, railroad dieselization) made it possible to squeeze progressively more output from a given unit of energy.

Nevertheless, the 1973–74 oil shock spawned an outpouring of writing, congressional testimony, and debate over the question of how much energy society "really needed" to support its industrial activity and personal comfort. Frequently, thermodynamic criteria were advanced as being more relevant than economic ones in answering the question. Witnesses to these arguments may recall the wide range of rhetoric evoked by the topic— from coolly analytical to hortatory, from restrained in tone to stridently polemical.

One influential figure around whom the critique of the supposedly "conventional" energy mindset coalesced was Amory Lovins. Combining detailed analysis with a vehement but sophisticated and effective advocacy, Lovins despaired of a supply-side bias toward big, "hard-path" energy facilities as opposed to a "soft-path" demand-side orientation. The latter emphasized eliminating waste and inefficiency in energy use. To Lovins, the energy problem was symptomatic of problems besetting other aspects of society: environmental threats, nuclear proliferation, economic distress, and—for good measure—dependence on the decisions and biases of a technological elite. His writings attracted a wide following and readership (Lovins, 1976; 1979).

Sensitive to the charge that his prescription for remedying defects in the energy system implied significant government intervention and controls, Lovins argued that, on the contrary, public policies that impair the functioning of competitive markets lie at the root of these problems. Preferential subsidies to nuclear and other energy systems crowd out soft-path alternatives. Inadequate penalties for environmental and other externalities, as well as capital market distortions, discourage energy conservation investments. There is also an imbalance between utility capacity enhancement and demand-side management strategies. If these market imperfections cannot easily be remedied, then government must compensate. For example, R&D support for solar energy is a justifiable antidote for nuclear favoritism.

Even those whose perspectives on energy and society were less sweeping than Lovins's recommended a more active government role in energy markets. In so doing, they implicitly dismissed the notion that the events that helped elevate the energy debate to such prominence in the first place—the new energy price regime—could bring about the conservationist response for which they pleaded. They particularly distrusted the economic construct of price elasticity. Energy price increases could not be relied upon to promote energy efficiency satisfactorily. That same message was a theme of another influential study of the 1970s, which, while acknowledging the impact of price on demand, also gave to government policies a powerful voice in promoting consumption or, more desirably, encouraging conservation (Ford Foundation, Energy Policy Project, 1974). In one of its scenarios the project postulated the feasibility, without economic penalty, of zero energy growth as a normal state of affairs as early as 1985, a course of development not dependent on the effects of the price shocks of 1973–74. While such a perspective raised questions about the authors' use of the term *feasibility*—zero energy growth clearly denoted some major lifestyle changes and other adjustments—like Lovins's work, this project succeeded in further integrating conservation and efficiency questions into contemporary energy debate.

Reprise of Policy Responses

While the most direct policy challenge arising from the 1973–74 oil crisis, reinforced by the events of 1979–80, was shielding the country from oil import disruptions, efforts were undertaken to complement that goal with other measures. Although many were destined to vanish within a brief number of years, a bewildering array of bureaucratic "alphabet soups" came into being. FEA (the Federal Energy Administration, itself a successor to the Federal Energy Office), prior to becoming DOE (the U.S. Department of Energy) in 1977, presided over one of the most notable federal energy forecasting exercises ever undertaken: the *Project Independence Report* of 1974. The background to this enormous undertaking, involving hundreds of people in and outside government, had been President Nixon's November 7, 1973, speech pledging that "by 1980, under Project Independence, we shall be able to meet America's energy needs from America's own energy resources" (transcript of presidential address, as quoted by Greenberger, 1983). Although the *Project Independence Report* shied away from trumpeting total self-sufficiency, it observed (U.S. FEA, 1974: 14) that by 1985 the nation "could" achieve zero imports with real oil prices remaining at the quadrupled postembargo level. Fifteen months after this report, an updated analysis (U.S. FEA, 1976) acknowledged that the unfolding domestic supply picture was more disappointing than earlier judged. However, because of price firmness and new legislation, the updated "FEA analysis shows that energy independence can still be achieved" (p. xxi).

Another member of the alphabet-soup family was ERDA (the Energy Research and Development Administration), established in 1974 to manage nonregulatory nuclear activities and energy research. In 1975 EPCA (the Energy Policy and Conservation Act) mandated automotive CAFE (corporate average fuel economy) standards, continued oil price controls, and created the SPR (Strategic Petroleum Reserve). The latter is an emergency stockpile targeted, at various times, at between 500 million and 1 billion barrels (its size at the end of 1992 was 575 million barrels). Its rationale and efficacy in reducing international oil price shocks remains in dispute today. In 1978 PURPA (the Public Utility Regulatory Policies Act) sought to promote innovative resource and technology applications to electric generation, as governed by avoided-cost criteria. NECPA (the National Energy Conservation Policy Act), also passed in 1978, required utilities to provide conservation services and introduced mandatory efficiency standards.

In the nonacronym category were three notable pieces of legislation, although the only one of lasting significance was the Natural Gas Policy Act of 1978, setting a schedule for decontrol of wellhead gas prices. Another was the (since repealed) Powerplant and Industrial Fuel Use Act, also enacted in 1978, which barred the use of oil and natural gas in new power plants immediately and in existing plants after 1990. Finally, the Energy Security Act of 1980, which created the quasi-independent (and short-lived) Synthetic Fuels Corporation, offered financial incentives to stimulate production of oil-shale- and coal-derived fuels.

This panorama depicts the intense pace of legislation caused by the energy upheavals of the 1970s. It does not trace how each of these policy departures played itself out. By the 1980s many of these efforts had ended, been amended, or become part of major institutional reorganizations. Also, no categorical judgment is possible on the extent to which each of the specific policy responses cited was appropriate or ill considered, and whether outcomes were wholly positive or negative. To this day, for example, there are respectable arguments on both sides as to the influence of obligatory CAFE standards as opposed to market forces in bringing about the impressive automotive fuel-economy gains of the last twenty years.

There is less doubt that PURPA stimulated innovative energy use, notwithstanding the cost and economic distortions it produced in its early years. It is also important to leaven ex post criticism with a dose of charity: even the synfuels program, now widely seen as a symbol of governmental hubris, contained at least some reasonable elements (e.g., it included a basic research component in its design phase). Finally, even questionable programs and policies—such as price controls, now seen to have been monstrous mistakes—deserve some grudging respect if they help us avoid similar future missteps.

The Record

Even as new policies were proposed, statutes enacted, and ideologies pressed, domestic and international energy patterns and markets registered significant changes.

U.S. Aggregate Energy Price and Consumption

Figures 2.1 and 2.2 depict several key indicators that leave little doubt about the impact of the price escalation of the 1970s. During the quarter-century preceding the Arab oil embargo of 1973–74, real energy prices in the United States—measured as a composite at the primary consumption stage—declined by about 1 percent annually (fig. 2.1). With the additional stimulus provided by growth in real income, per capita energy consumption rose by 2 percent per year. The sharp rise in energy prices into the early 1980s was accompanied by significant declines in per capita energy consumption and moderate declines in overall consumption (the left-hand axis of fig. 2.2). The decline became particularly marked with the second oil price shock, no doubt reinforced by the turnover of energy-using capital stock—first set in motion by the earlier price hike. After the early to mid-1980s, real prices once again fell while consumption increased. By 1990 the real price index

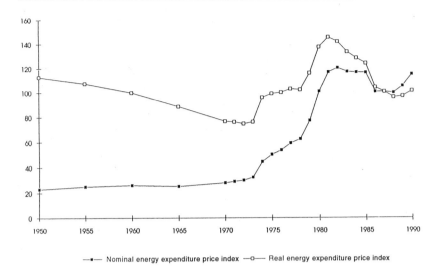

Figure 2.1   Index Numbers of U.S. Energy Prices, 1950–1990 (1987 = 100)
Sources: 1950–70 series adapted from Schurr et al., 1979; 1970–90 series from U.S.
DOE/EIA, 1993.
Notes: The 1950–70 series was linked to the 1970–90 series at 1970. Note that for the
1950–70 period, only the five-year benchmarks were plotted.

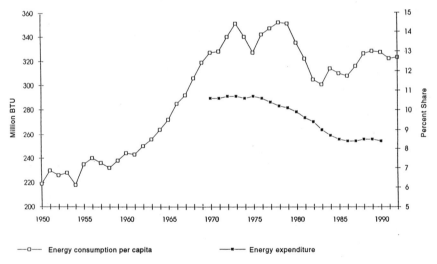

Figure 2.2   Two Alternative Measures of U.S. Energy Consumption, 1950–1992:
Energy Consumption per Capita (millions of BTU) and Energy Expenditure (per-
cent of gross domestic purchases, 1987 prices)
Source: U.S. DOE/EIA, 1993.

stood at a level 70 percent below the peak reached after the post–Iranian Revolution price spike. Reflecting these trends is the plot of energy expenditures as a share of gross domestic purchases (the right-hand axis of fig. 2.2): sharply declining between the early 1970s and the mid-1980s, then leveling off.

Evidence pointing to the strongly deterministic influence of price on these trends is overwhelming. Hogan (1984: 35) states, "We need make no appeal to structural change in the aggregate data: the period before the oil shocks is consistent with the decade of turmoil after the oil embargo. And we need look no further than a simple economic explanation of energy choices: prices and economic activity can explain the movements in energy demand within a framework of slow adaptation driven by the turnover of the capital stock of energy-using equipment." Findings by Hirst and others (1983), who also dissected reasons for the slowdown in growth of U.S. energy consumption between 1973 and 1981, further confirm these trends. Adjusting for the effect of a slowdown in economic growth, these authors attribute two-thirds of the reduction below trends extrapolated from pre-embargo conditions to rising fuel prices.

And yet the government's 1970s energy projections gave far too little play to demand elasticity. A May 1979 DOE report projected that world oil prices would reach a 1990 level of $28/barrel (1981 prices—which translates into $40/barrel in 1990 prices). U.S. energy consumption was projected to reach 102 quadrillion BTUs ("quads") in that year. In fact, 1990 oil prices stood at half the projected level, and U.S. energy consumption stood at just 81 quads. Since the average annual GDP growth rate of 2.6 percent realized during the decade was not appreciably below the projected rate of 2.8 percent, faulty elasticities rather than faulty economic growth assumptions can be blamed for the flawed projections (Second National Energy Plan of May 1979, cited in U.S. DOE, 1991, table 1-6).

Energy, Oil, and Electricity Consumption

Figures 2.3 and 2.4 plot long-term (pre- and postembargo) trends in overall U.S. energy consumption, all oil products, gasoline, and electricity. Each series shows a conspicuous break in trend during the 1970s. Unlike the other categories plotted, electricity largely sustained its *absolute* growth, albeit at a greatly decelerated pace. This was a source of much financial discomfiture to utilities engaged in costly nuclear and other construction projects during the 1980s predicated on past growth trends (i.e., those before the oil shocks). Thus, projects planned in the 1960s and expected to come on line a

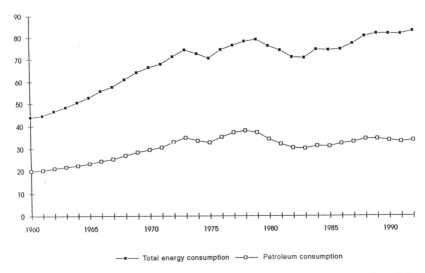

Figure 2.3 U.S. Energy Consumption, Total and Petroleum, 1960–1992 (quadrillions of BTU)
Source: U.S. DOE/EIA, 1993.

decade or so later assumed the roughly 7 percent annual growth rate (a decadal doubling) that had been the historic norm. The middle panel of figure 2.4 shows the conspicuous slippage from that extrapolated path in actual development. The lower panel records improvements in automotive fuel efficiency, a key factor in flat gasoline consumption.

Energy Mix Changes

Figures 2.5 and 2.6 depict shifts in the composition of primary energy sources for the United States and the world. Common to both is the relative decline of oil and the rising proportion of nuclear power. Coal increased its share in the United States. Stimulated by significant discoveries in Western Europe and other regions, natural gas raised its worldwide share. Coal has historically accounted for—and it continues to represent—a comparatively high proportion of global energy, while the same has been true of natural gas in the United States. Although both coal and nuclear power raised their proportionate share of U.S. energy, their actual performance did not live up to forecasts of a decade earlier. U.S. nuclear generating capacity was projected to reach over 150,000 megawatts (MW) by 1985, producing over 25 percent of the nation's electricity (U.S. FEA, 1976). For well-known reasons arising from safety and management concerns, cost overruns and debt burdens, and lower-than-predicted electricity demand, that scenario failed

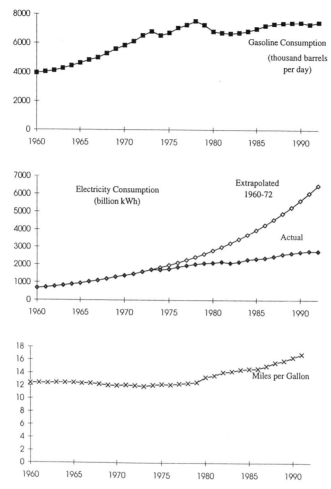

Figure 2.4   **Measures of U.S. Electricity and Gasoline Consumption, 1960–1992**
Source: U.S. DOE/EIA, 1993.

to ensue by 1993, when nuclear capacity stood at only 100,000 MW. That capacity level, however, furnished 22 percent of the nation's electricity use. Coal production, in part benefiting from the expected demands of a growing synfuels industry, had been projected to reach 1.04 billion short tons by 1985, a figure that has begun to be approached in only the last couple of years.

U.S. Oil Import Share

Notwithstanding the leveling off in U.S. petroleum consumption, the share of imports has remained high (see fig. 2.7) and in recent years has been

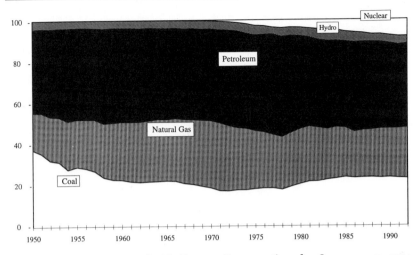

Figure 2.5   Composition of U.S. Energy Consumption, by Source, 1950–1992 (percent)

Source: U.S. DOE/EIA, 1993.

Note: Hydro includes geothermal and other.

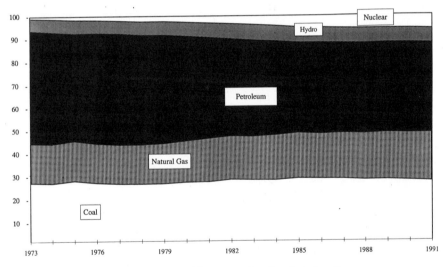

Figure 2.6   Composition of World Energy Production, by Source, 1973–1991 (percent)

Source: U.S. DOE/EIA, 1993.

Note: Petroleum includes natural gas plant liquids.

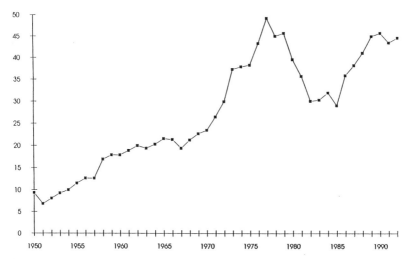

Figure 2.7   Net Petroleum Imports as a Share of U.S. Petroleum Consumption, 1950–1992 (percent)
Source: U.S. DOE/EIA, 1993.

edging up to compensate for the decline in U.S. oil production. Early post-embargo scenarios of prospective oil import levels and consumption shares are among the more interesting postmortems in which one can indulge. There was a strong element of political posturing in President Nixon's and Project Independence's visions of the possibility of zero imports within the decade. But beyond that there were overly optimistic judgments about responses on the domestic oil supply side.[4] Project Independence failed to foresee a 1985 level of domestic oil production any lower than 13–17 million barrels per day instead of the constant 11 million barrels per day that materialized. With actual 1985 prices considerably higher than those assumed, the supply side of the analysis turned out to be flawed. Thus, notwithstanding the prolonged plateau in U.S. oil consumption, imports dropped from 6.3 million barrels per day in 1973 to 5.1 million barrels per day in 1985.

OPEC's Share of World Oil Production

OPEC's dominance grew throughout the pre-embargo decade, with the cartel's share of world oil output rising from 42 percent in 1963 to the 56 percent peak recorded in 1973. Amid the slowdown in consumption and expanding output from the North Sea and other areas, OPEC's subsequent decline was precipitous, its share reaching 31 percent in 1985. OPEC's recovery, to a 42 percent share in 1992, stems from the former Soviet Union's

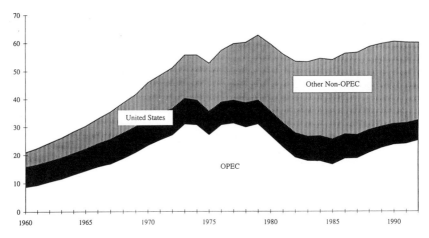

Figure 2.8 World Oil Output, by Major Producing Region, Absolute Numbers, 1960–1992 (millions of barrels per day)
Source: U.S. DOE/EIA, 1993.
Note: Crude oil production excludes natural gas plant liquids.

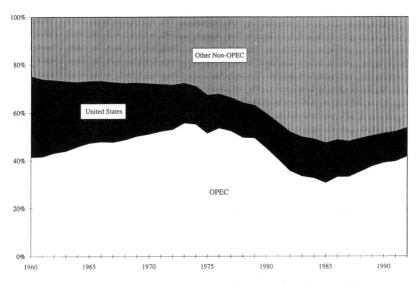

Figure 2.9 World Oil Output, by Major Producing Region, Percent Shares, 1960– 1992
Source: U.S. DOE/EIA, 1993.
Note: Crude oil production excludes natural gas plant liquids.

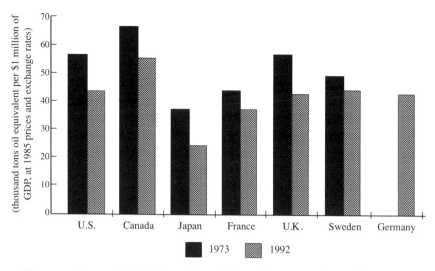

Figure 2.10   **Energy/GDP Ratios, Selected Countries, 1973 and 1992 (thousands of tons of oil equivalent per $1 million of GDP, at 1985 prices and exchange rates)**
Source: IEA, 1993.
Note: The numerator refers to the OECD's measure of energy consumption—namely, "total primary energy supply." No data provided to OECD for Germany, 1973.

collapsing export performance, a continuing decline in U.S. production, and depressed prices since 1985, which have probably discouraged some non-OPEC production. These trends are plotted in figures 2.8 and 2.9.

Energy and GDP

Differences in industrial structure, housing patterns, and other factors lead to variations in aggregate energy-intensity among countries. For example, Japan's post-1973 decision to import aluminum metal rather than continue operating domestic smelters "improved" the country's energy-intensity standing at the expense of the exporting countries' "deterioration." To be sure, differential energy taxes—such as those on motor fuel, adopted decades ago, in most cases for reasons unrelated to energy conservation or the environment—prompt consumers to economize their energy use. But it would be misleading to interpret international variability in energy/GDP ratios (as shown in fig. 2.10) as a valid indicator of either disciplined or profligate energy use. The energy price shocks of the 1970s undoubtedly elicited behavioral responses in industrial processes, buildings, and transport that contributed to the generally declining energy/GDP ratios for countries shown in figure 2.10. As yet there is no evidence of significant convergence in ratios among countries, although until the oil price collapse in the 1980s, the United States reduced its energy/GDP ratio at a faster pace than almost all twenty-three other OECD countries.

Lessons Learned and Not Learned

Oil Imports and Energy Security

In some respects, oil import concerns have evolved a great deal since the first oil shock in 1973. In other respects, they have not changed at all. The initial reaction to the shocks was that the nation needed to insulate itself from the unstable world oil market by becoming self-sufficient. This extreme view, effectively refuted in the (ignored) findings of the Cabinet Task Force on Oil Import Control (1970), was quickly discarded in favor of the less costly alternative of reducing America's vulnerability to external shocks. Still, there are many ways to go about reducing vulnerability, with varying degrees of effectiveness and wide cost differences. The nation has never been very clear about the nature and magnitude of the problem or about its willingness to pay to reduce it. Nevertheless, concern about the magnitude of oil imports never goes away, and each successive administration has struggled to determine what to do about it.

Ultimately, the problem of oil imports hinges on the costs imposed on the entire economy when the price of oil surges during a market disruption. After the two oil price shocks of the 1970s, it was thought that these economic costs must be very high. In both cases the United States suffered deep recessions, with sharp and immediate reductions in GDP and corresponding increases in unemployment. Moreover, a number of other industrial countries had the same experience. Consequently, it was widely presumed that oil price increases were damaging to industrial economies, and that oil-importing nations should develop the means to prevent oil price shocks and to reduce their economic vulnerability to price change (Bohi, 1989).

At first it was thought that energy security could be improved merely by reducing oil imports. This view was called into question when it was observed that in 1979–80 the United Kingdom experienced the deepest recession among the major industrial countries despite becoming an oil exporter, while Japan avoided a recession altogether despite complete dependence on imports for all its energy needs. Analysts began to realize that adjustments forced by changes in the world price of oil are no different when the domestic oil price changes. And since domestic oil price is determined by world price, solely reducing imports would not improve energy security.

This realization put greater importance on the objective of stabilizing the world price of oil during crises. This objective was the primary purpose of the International Energy Agency. The United States responded to the challenge by building a Strategic Petroleum Reserve with a capacity now close to 600 million barrels, at a cost in excess of $20 billion. With two exceptions the SPR has never been used. A test sale was made in the fall of 1990 in the wake of the Iraqi invasion of Kuwait. And in January 1991, at the beginning of Operation Desert Storm, an actual sale took place that elicited little interest. All in all, the presence of the SPR might be viewed as having a stabilizing effect on the oil market because oil traders believe that the release of oil might affect their speculative profits.

The most important unresolved energy security issue is the vulnerability of the economy to energy price shocks. The economic impact of price shocks is thought to arise from the adjustment costs that reverberate throughout the economy in response to changes in the price of energy relative to that of other inputs into the production of goods and services. The cost of adjusting to higher energy prices arises because of price and wage rigidities in the economy that prevent the smooth accommodation to higher oil prices without losses in output and employment. Consider wage rigidities, for example. It is thought that higher energy prices reduce demand for labor because labor and energy are used as complements in the production of other goods and services. When the price of energy rises, less energy is used and the productivity of labor consequently falls. A decline in labor productivity means that labor costs have increased, and employers will attempt to lower these costs by reducing wages or employment. If wages are rigid, because of constraints such as labor contracts or annual wage adjustments, then employers have no choice but to reduce employment, thereby diminishing the aggregate performance of the economy.

While this is a very plausible explanation of energy vulnerability, the importance of the effect of energy prices on the economy has not been adequately determined. Studies that focus on the correlation between higher energy prices and recessions suggest that the cost is very high, although these studies usually ignore the fact that Japan somehow managed to avoid a recession in 1979–80, and that none of the industrial countries experienced an economic boom when the price of oil fell by half at the beginning of 1986. Studies that look beyond aggregate output and associate economic performance (by disaggregated industrial sector) with energy-intensity find that energy prices appear to have little to do with performance. An alternative explanation for recessions is that monetary authorities, rather than conducting an expansionary monetary policy—which, by reducing real wages, lowers labor costs and mitigates the effect of rigid wages on the economy—tighten the money supply in response to higher oil prices, thereby driving the economy into recession.

If the economy is already experiencing rapid inflation, it may not be feasible to implement an expansionary monetary policy—which in principle could mitigate the outflow of payments for oil imports—to accommodate higher energy prices. This was the situation during the 1970s. It may also be impractical to make wages less rigid. This would involve major changes in social institutions. Nor is it certain that the effort would be worth the cost. We simply do not know enough about the importance to the economy of changes in the relative price of energy. The range of options for energy security is effectively narrowed to encouraging technological improvements and behavior that increase the elasticity of supply and demand for energy, and for oil in particular. On the consumption side, for example, "no-regrets" policies designed to meet environmental goals through pollution taxes would be one means of bringing about positive spillover benefits for energy security from more efficient and environmentally benign energy use.

This conclusion departs widely from beliefs and actions of twenty years ago. It means that the best course for energy policy is one that relies on the market to maximize efficiency, and that seeks to enhance market efficiency by expanding the range of energy options. Because no country can isolate itself from the energy security problem, it behooves every country to cooperate in improving market efficiency.

The Role of Government

The oil price shocks of the 1970s, whatever their cause and cost, produced dramatic changes in fundamental views of economic behavior and the role of government in the market. The initial response to the oil crisis, following a well-established, decades-old tradition, was more government intervention in energy markets. It took several years for the realization to take hold that earlier government regulation was responsible for much of the crisis. It also took several years to realize that more government interference would not solve anything. By 1978 it became clear even to an intervention-prone Congress that existing trends had to be reversed and that only by moving toward less regulated markets could economic waste and inefficiency be reduced. Legislation was passed in that year that committed the nation to deregulating oil and natural gas prices, and began a process of reducing regulation of natural gas pipelines and electric utilities that continues to this day.

While the benefits of deregulation have been too obvious to generate much enthusiasm for a return to the pervasive forms of intervention that preceded and immediately followed the first oil price shock, the illusion survives that a neat governmental policy fix can spare us the consequences of energy markets' uncertainty. An example is proposals for indefinitely extending the ban on exporting North Slope oil. Oblivious to the price controls and allocation schemes responsible for the gasoline lines of the 1970s, and to our limited capacity to influence the world market price of oil, a supporter of that extension, Senator Patty Murray of Washington, recently commented, "Alaska North Slope oil provides an insurance policy to consumers on the West Coast that the giant gasoline lines of yesterday will not reappear because of the irrational acts of some Middle East despot or a group of crazed religious zealots" (Environmental and Energy Study Institute, 1994: B3).

The U.S. Role in the World Oil Market

At the time of the first oil shock, it was widely thought that the United States could effectively isolate itself from world oil market instability. This is evident in President Nixon's Project Independence, which had complete autarky as its objective. We have since learned that the domestic economy cannot be shielded from events in the world oil market irrespective of how much oil we import. A succinct, if belated, acknowledgment appears in the government's *National Energy Strategy* (U.S. DOE, 1991: 3). This document recognizes that domestic oil prices are determined by world oil prices. It also acknowledges that the United States, notwithstanding its quantitatively large role, cannot influence the world oil market without taking into account the actions of the rest of the world. An increase in oil demand, whether originating here, in Europe, or in Asia, has the same effect on the world oil market, and an increase in oil supply has the same effect no matter where it originates.

At the same time, the behavior of the world oil market is the product of its individual components. World demand for oil is more elastic if consumers worldwide have more alternatives to oil to satisfy energy demands. The world supply of oil is more elastic if there are more producing regions that can enter the market at higher prices. And the actions of any group of producers have less of an effect on world price if oil demand and/or supply is more elastic. The interdependent character of the oil market also means that policies implemented by the United States alone will have limited effects on the oil market and could be offset by other countries' actions. Thus, releases of oil from the SPR will be small relative to the size of the oil market, and the effect could be overwhelmed by increases in consumption or stockbuilding in other countries.

Thus, actions by OPEC to control price must have the cooperation of many large oil producers. Likewise, efforts by oil-consuming countries to stabilize price must have widespread cooperation among major consuming countries.

Strengths and Weaknesses in Energy Analyses

Experience with the energy price shocks of the 1970s taught many lessons about economic behavior and about the limited ability of governments to alter that behavior, particularly when market forces run counter to policy prescriptions. That experience reminds us that people and firms do respond to economic incentives by altering their energy consumption, by making investments in new technology, and by reshaping or strengthening the basic institutions involved in international energy trade.

The events of the 1970s could not be repeated in the world of today. Energy markets work in fundamentally different ways than they did two decades ago. In general, the difference may be described in terms of the petroleum market. Oil is now traded much like any other commodity, with active spot and futures markets, anonymous transactions at open prices, and instantaneous responses to changes in market conditions. These characteristics did not obtain twenty years ago.

Another oil crisis may occur tomorrow and may be just as severe as those experienced in the past, but the next crisis would have to originate and develop differently. For example, demand cannot outpace supply for very long without a price response, and oil traders will not engage in stockbuilding and speculation without considering the risk of losses from a downward movement in price. It is to be hoped that governments will also learn from past mistakes and refrain from imposing price and allocation controls to manage the next crisis. If the past has taught us anything, it is that markets manage crises much better than governments do.

Despite the broader lessons of previous experience, our knowledge of some key analytical details remains surprisingly poor. Three examples illustrate these shortcomings: international oil price behavior; elasticities of demand and supply for energy; and the impact on the economy of changes in energy prices. With the benefit of hindsight, we know only in very general terms why the price of oil has behaved as it has over the past twenty years. Especially important to the oil price shocks of 1973–74 and 1979–80 was the increase in inventory demand that occurred around the world in response to political events and market conditions. However, the importance of inventory behavior in the two price shocks was not appreciated when it occurred, nor is it understood to this day why stocks were built up as they were and why prices responded as they did. One reason for the lack of understanding is the lack of inventory data beyond the stocks held by oil refiners in OECD countries. For this and other reasons, no one has adequately explained why the price of oil rose as high as it did after 1973, why it remained so high until 1982, and precisely what role OPEC played in the outcome. While the common perception is that OPEC engineered the whole thing, the facts do not support that perception.

Another reason for the generally poor understanding of oil price behavior was poor understanding of the responsiveness of oil supply and demand to price changes. A view commonly held before the oil shocks of the 1970s, and for a few years after, was that U.S. energy consumption was unresponsive to price increases. It was thought that the American driver could not be persuaded to give up the large American-made car in favor of smaller imported models. American car manufacturers, subscribing to the same myth, continued to produce large gas guzzlers and maintained production long after imports of smaller and more efficient models had encroached on American market share. To some extent these rigid views may be traced back to the long history of stable gasoline prices. The price was low and stable for so long that it was easy to develop the impression that price did not matter in consumption decisions. Econometric studies of gasoline demand, relying on stable price experience, could not dispute this impression: the magnitude of price variance was too small to reveal any correlation with gasoline consumption. Hence, price was viewed as unimportant before the oil price shocks, and it was expected to remain unimportant afterward.

The post–oil shock experience of the 1970s pointed unmistakably to different phenomena at work. The significance of the elasticity of oil demand *and* the time required for the price to take effect are evident in figure 2.3. Petroleum consumption in the United States actually rose for three years after the 1973–74 oil shock but then fell after 1978. It was not until 1989 that U.S. petroleum consumption matched the 1973 level.

There was a similar experience in consumption of coal, natural gas, and electric power. The latter case is particularly instructive because, as noted earlier, it demonstrates the large costs incurred by society when consumption forecasts are consistently and systematically wrong. Electricity prices were considered so unimportant to electricity demand before the 1970s that utility planners did not factor price into demand forecasts when projecting capacity requirements. Consumption had risen at an annual rate of 7 percent for years, and investment in new electric generating plants, including expensive new nuclear plants, grew to keep pace. Because of the long lead time in building new plants, moreover, construction had to be well under way before the increase in consumption was realized. Thus, when the oil price shocks drove up all energy prices, the electric industry had a large construction program under way, and it believed that consumption would grow to absorb the new capacity.

The experience also put to rest the belief that electricity was not susceptible to the economic turmoil that affected oil. Fuel costs account for about 40 percent of the cost of generating electricity, but they are not the whole story. Electric utilities are capital-intensive firms. With inflation and rising interest rates forcing construction costs well above planned levels, new capacity came on line well over budget. This happened at the same time that consumption growth collapsed. In 1974, for the first time in memory, electricity consumption actually declined. Thus, expensive new capacity no longer had a market to serve, and public utility regulators began to blame the utilities for poor planning. A new chapter in the history of the industry was written in the 1970s and 1980s, when regulators began refusing to pass along higher costs to consumers and forced utility shareholders to absorb some of the losses. This played a major role in the transformation of the electric utility industry into one that is less rigidly regulated and more competitive than in the past.

The quality of information about demand elasticities has improved during the last two decades and is substantially better than that for supply. Even so, it still can be judged adequate only for some energy products consumed by some consumer classes. For example, our understanding of the behavior of residential demand for electricity in the United States is now pretty good, while our insight into what drives commercial and industrial demand for energy remains poor. For most energy products in most countries, very little is known about consumption. The lack of information is particularly glaring outside of the most advanced industrial countries. Since total world oil demand is the sum of the contributions from all individual countries, information about all of the parts is necessary to understand the behavior of the whole.

Finally, it must be borne in mind that we have only a weak understanding of the effect on the economy of changes in the price of energy. Basically, the analytical problem involved in understanding this relationship is one of separating the influence of energy prices from that of other factors operating on the economy, some of which could overwhelm the importance of energy. To date, this has not been done. Consequently, those advocating one position or another on this issue are operating largely on the basis of ignorance. Nevertheless, large sums of taxpayer money have been spent, and still larger sums proposed, for programs based on the premise that the economic cost of higher energy prices is very high.

The Social Cost of Energy

What does the experience of the last two decades suggest about our ability to manage the social and environmental problems associated with energy? The recognition that energy production and use expose a significant market failure—a divergence between a private benefit-cost reckoning and the broader benefit-cost calculations that include environmental and other external effects—was hardly a discovery attributable to the energy upheavals of the 1970s. But the wide-ranging policy debates and analyses set off by those events have brought that recognition more sharply into public consciousness and helped to provide the political consensus needed for significant new energy-environmental initiatives. There has also been a growing consensus among interest groups that when unconstrained private actions harm the environment, marketlike instruments are likely to promote environmental quality with greater flexibility and at lower cost than rigid government mandates. The SO_2 tradable-permits provisions of the Clean Air Act Amendments of 1990 are a case in point.

The United States can take legitimate pride in having moved a number of such environmental quality management tools to a stage of development beyond that in other industrial countries, not to mention the developing world. For example, in Venezuela, energy markets are rigged through perverse control of gasoline prices at substantially less than half the world level. While such a policy might be understandable in the light of domestic populist considerations, it is oblivious to social costs.

Tolerance for some limited reshaping of energy market signals to reflect externalities has also benefited from the fact that, with a 5 percent or less share of GDP, energy is not so central a driver of economic conditions as some industry groups suggested twenty years ago. This has provided latitude to pursue at least limited social/environmental goals without anxiety that this inevitably poses a zero-sum trade-off between the environment and economic well-being.

But this argument should not be carried too far. Notwithstanding our ability to achieve better environmental outcomes through reliance on marketlike instruments for well-recognized, regionally identifiable problems (e.g., local air pollution), successfully managing long-term *global* threats is going to be far more difficult. Thus, with respect to global warming, the science is uncertain and in dispute; time horizons are well outside the planning range of business, households, and government; and multinational approaches (e.g., CO_2 emissions trading), though conceptually defensible, are politically unappealing. Not only would sovereign nations have to submit to untested international ground rules, but there appears to be a deep-seated reluctance to bow to environmental policies fashioned by the wealthy countries that created the environmental problems in the first place.

The climate-change threat—and, more broadly, the goal of long-term "sustainability," which in some discussions has become a facile, mindless catchword—poses a more intractable externality dilemma than the problems on which we have made reasonable progress. Nonetheless, the sustainability debate points up serious and legitimate questions, which neither economists nor ecologists are close to resolving: To what extent does the present generation have a responsibility, in managing resources, to future generations? By what means should this responsibility be exercised? Here, *resources* takes on broad scope: not only physical resources like energy, but species, ecosystems, and numerous other valued elements of the natural system. Ideally, having learned some important lessons from the last twenty years puts us in a better position to address the problems of the next two hundred.

Déjà Vu?

Lest one conclude that the lessons of the last two decades have elevated the "do-no-harm" test of public policy to greater prominence, the news is not auspicious. Consider, as a particularly egregious example, federal rules on diesel fuel sales that became effective January 1, 1994 (*New York Times*, 1994: A1, 16). These legislatively mandated regulations seek to meld an uncoordinated set of energy and environmental statutes: imposing taxes inversely related to sulfur content; allowing the American Red Cross, but not the Salvation Army, to purchase the fuel tax-free; and differentiating among clear, red-, and blue-tinted versions of the product according to uses and tax treatment. Thus, if you're driving a museum vehicle, you can fill up with red diesel, but if you're operating antique aircraft, it's blue kerosene for you. Trucks must use the taxed, clear liquid for transport but can use the untaxed blue variety for refrigeration. These new rules went into force *after* the "reinventing-government" exercise. But then, even the penal system doesn't show a 100 percent success rate with rehabilitation. There's still hope.

NOTES

We are grateful to Mary Elizabeth Calhoon for research assistance in the preparation of this chapter and to Vito Stagliano for helpful comments on an earlier draft.

1. We digress with a personal aside about the intellectual schisms of those years. The RFF volume *Energy in America's Future* (Schurr et al., 1979: 93) contained a table that mistakenly showed real electricity prices increasing by 18.3 percent between the *pre-embargo* years 1971 and 1972 rather than (as corrected in a subsequent printing of the book) by 1.7 percent. It may be harsh, in retrospect, to suggest that such an extraordinary price rise might have aroused researchers' suspicions. Be that as it may, in a letter to *Science* magazine John Holdren, Paul and Anne Ehrlich, and John Harte (1980: 1296–1301) seized upon that apparent price spike to argue that energy scarcity had manifested itself as a phenomenon *prior to and apart from* OPEC's exercise of market power. Julian Simon—whose earlier *Science* article (1980b) Holdren and his co-authors were critiquing—rebutted their letter in the same issue with barely concealed glee by noting the fact of the typo and its crucial importance in his critics' argument (Simon, 1980a: 1305–8). He went on to observe that even "if correct, it would seem scientifically imprudent to rely for any general conclusion upon a single number against the contrary evidence of a sweep of data three-quarters of a century long that tells a radically different story." So much for the passions that, even in the academic community, energy and natural resource issues can evoke.

2. Given the 60 percent increase in the general price level between 1974 and 1980, this second oil price shock involved a real escalation over pre-embargo levels of closer to 25 percent than to 100 percent.

3. For example, the Paley Commission cited a National Petroleum Council estimate "that gasoline made from shale on the basis of a 200,000-barrel-per-day operation could be sold in the Los Angeles market at 14.7 cents a gallon . . . compared with a contemporary Los Angeles price of 12.7 cents on gasoline from crude in October 1951" (President's Materials Policy Commission, 1952: 39).

4. Paradoxically, the underestimation of demand-side responses to higher prices was accompanied by *over*estimation on the oil supply side. That is, it was expected that higher prices would stimulate a significant rise in U.S. production.

REFERENCES

Bohi, Douglas R. 1989. *Energy Price Shocks and Macroeconomic Performance.* Washington, D.C.: Resources for the Future.

Bohi, Douglas R., and Milton Russell. 1978. *Limiting Oil Imports: An Economic History and Analysis.* Baltimore: Johns Hopkins University Press, for Resources for the Future.

Boulding, Kenneth E. 1972. "New Goals for Society?" In *Energy, Economic Growth, and the Environment,* edited by Sam H. Schurr. Baltimore: Johns Hopkins University Press, for Resources for the Future.

Cabinet Task Force on Oil Import Control. 1970. *The Oil Import Question: A Report on the Relationship of Oil Imports to the National Security.* Washington, D.C., February.

Commoner, Barry. 1972. "The Environmental Cost of Economic Growth." In *Energy, Economic Growth, and the Environment,* edited by Sam H. Schurr. Baltimore: Johns Hopkins University Press, for Resources for the Future.

Environmental and Energy Study Institute. 1994. *Weekly Bulletin.* Washington, D.C.: Environmental and Energy Study Institute, March 14.

Exxon Corporation. 1981. *Exxon USA.* New York: Exxon, third quarter.

Ford Foundation, Energy Policy Project. 1974. *A Time to Choose: America's Energy Future.* Cambridge, Mass.: Ballinger.

Greenberger, Martin. 1983. *Caught Unawares: The Energy Decade in Retrospect.* Cambridge, Mass.: Ballinger.

Hirst, Eric, Robert Marlay, David Greene, and Richard Barnes. 1983. "Recent Changes in U.S. Energy Consumption: What Happened and Why." *Annual Review of Energy* 8.

Hogan, William W. 1984. "Patterns of Energy Use." Discussion Paper Series E-84-04. Energy and Environment Policy Center, Harvard University, Cambridge, May.

Holdren, John P., Paul R. and Anne H. Ehrlich, and John Harte. 1980. Letter to *Science,* December 19, pp. 1431–37.

International Energy Agency (IEA). 1993. *Energy Policies of IEA Countries: 1992 Review.* Paris: Organization for Economic Cooperation and Development.

Kirkland, Richard I., Jr. 1984. "Exxon Rededicates Itself to Oil." *Fortune,* July 23, p. 31.

Lovins, Amory. 1976. "Energy Strategy: The Road Not Taken." *Foreign Affairs,* October, pp. 65–96.

———. 1979. *Soft Energy Paths: Toward a Durable Peace.* New York: Harper Colophon Books.

Massachusetts Institute of Technology (MIT). 1970. *Man's Impact on the Global Environment.* Report of the Study of Critical Environmental Problems. Cambridge: MIT Press.

———. 1971. *Inadvertent Climate Modification.* Report of the Study of Man's Impact on Climate. Cambridge: MIT Press.

Meadows, Dennis, et al. 1972. *The Limits to Growth.* New York: Universe Books.

New York Times. 1994. "What's Red, Blue, or Clear in the Diesel Tank?" January 9, pp. A1, A16.

President's Materials Policy Commission ("Paley Commission"). 1952. *Resources for Freedom.* Summary of vol. 1. Washington, D.C.

Schurr, Sam H., and Bruce C. Netschert. 1960. *Energy in the American Economy, 1850–1975: An Economic Study of Its History and Prospects.* Baltimore: Johns Hopkins Press.

Schurr, Sam H., et al. 1979. *Energy in America's Future: The Choices before Us.* Baltimore: Johns Hopkins University Press, for Resources for the Future.

Simon, Julian L. 1980a. Letter to *Science,* December 19, pp. 1305–8.

——. 1980b. "Resources, Population, Environment: An Oversupply of False Bad News." *Science,* June 27, pp. 1431–37.

U.S. Congress, House of Representatives. 1974. "Global Scarcities in an Interdependent World." Hearings, Committee on Foreign Affairs, Subcommittee on Foreign Economic Policy, 93d Congress, 2d session.

——. 1976. "United States Commodity Policies." Hearings, Committee on International Relations, Subcommittees on International Resources, Food, and Energy; International Economic Policy; International Organizations; and International Trade and Commerce, 94th Congress, 2d session.

U.S. Department of Energy (U.S. DOE). 1991. *National Energy Strategy.* Washington, D.C., February.

U.S. Department of Energy, Energy Information Administration (U.S. DOE/EIA). 1993. *Annual Energy Review 1992.* Washington, D.C., June.

U.S. Department of Energy, Office of Policy, Planning and Analysis (U.S. DOE/ OPPA. 1981. *Energy Projections to the Year 2000.* DOE/PE-0029. Washington, D.C., July.

U.S. Federal Energy Administration (U.S. FEA). 1974. *Project Independence Report.* Washington, D.C., November.

——. 1976. *National Energy Outlook.* Washington, D.C., February.

3. Energy Policies, R&D, and Public Policy

A. DENNY ELLERMAN

Four points developed by Doug Bohi and Joel Darmstadter in chapter 2 deserve further examination. First, the mechanism establishing international oil prices is not as puzzling as their analysis suggests. Second, they pay too little attention to the domestic component of the general increase in energy prices and resulting improvements to energy efficiency. Third, they neglect a major part of the policy response: energy R&D. Finally, the most lasting effect of the nation's experience with energy policy in the 1970s may turn out to be its contribution to changing opinion about the role of government and public policy.

International Oil Price Behavior

As Bohi and Darmstadter note, two aspects of crude oil market behavior appear to have been insufficiently appreciated in the 1970s: the role of inventory behavior, and the difference between short- and longer-run price elasticities. The mechanisms by which crude oil prices rose and were maintained, and OPEC's role in that process, appear reasonably clear. It is instructive to compare the market disturbances of 1973–74 and 1979–80 with those of the Persian Gulf War of 1990–91.

Significant increases in inventory demand occurred in 1973–74, 1979–81, and 1990–91, corresponding to Middle East disturbances that threatened the continuity of the major source of crude oil to the world market. The rising demand for inventory during these episodes was entirely rational. Uncertainty about supply increases the demand for a commodity that has no immediate, readily available substitute. What is harder to understand in the case of oil is the magnitude of the increases in demand and price. The explanation is important since the price caused by the momentary demand for inventory provided the reference point for the administered prices OPEC producers were able to sustain for many years.

Two factors contributed to the confusion surrounding crude oil pricing in these years. The first was poor understanding of inventory behavior in general and crude oil inventory behavior in particular. The events of 1990–91 demonstrated in fairly pure form the relationship between Middle East disturbances, inventory demand, and oil prices. These events demonstrated that near-term prices could increase dramatically and to a far greater degree than prices for more distant contracts. In 1990, spot and near-term futures prices more than doubled in the months immediately following the disturbance, while eighteen-month futures increased by less than 50 percent (Verleger, 1993: 48).

When markets are in backwardation, as is typical when unusual inventory demand is present, the difference in short- and long-term elasticities is reflected in explicit prices. When most crude was produced and sold under vertically integrated contractual arrangements—as it was in the 1970s, but not in 1990—it was probably easier, if foolish, to mistake the temporary price appearing in the spot market as an indicator of long-term value, and thus as an appropriate marker for establishing administered prices. One wonders whether backwardation would have been observed had futures markets existed in the 1970s, and, if so, which term price would have been taken as the appropriate reference price.

The second explanation for the price increases of the 1970s is that the Middle East was not the only, or even primary, source of supply uncertainty. Another was the response of consumer governments, which then heavily regulated the pricing and allocation of crude oil and petroleum products. The added effect of regulatory uncertainty is apparent in the topping-off of gas tanks. This phenomenon contributed to inventory demand in 1973–74 and 1979–80, because price and allocation controls called into question the availability of supplies when needed by customers. In 1990 the availability of gasoline at some price was never in doubt, and consequently this addition to inventory demand did not occur.

The critical difference between the 1990 disturbance and the earlier two was OPEC's response. Whatever the motive, in 1973–74 and again in 1979–80, OPEC adopted as "official selling prices" those observed in spot markets. Once adopted, the administered prices persisted long after the inventory demand that gave rise to them had dissipated.

The mechanism adopted by OPEC, the official selling price (OSP), was new only in form. It was an obvious extension of the posted price system in effect during the 1960s and early 1970s by which producer-country tax revenues were determined. The difference after 1973 was that producing governments were formally selling crude at the OSP, instead of collecting taxes on sales by concessionaires at assumed posted prices. Increases in tax rates and posted prices effected prior to 1973 yielded higher revenues with imperceptible effect on demand. Thus, it was not surprising that the 1973–74 price increases initially appeared to be similar. The events of 1979–80 were largely a repetition of what then appeared to be a happy formula for producers. Only later did the formula come to be tested, with well-known results. By 1990 there was no thought of establishing a higher price at what the market, in response to yet another Middle East disturbance, demonstrated to be the short-term clearing price. The lesson of long-term price elasticity was learned too painfully to bear repetition.

The Domestic Component of Price Increases

One enduring effect of the energy upheavals of the 1970s is the significant reduction in energy-intensity, or energy use per unit of GDP, illustrated in figure 3.1. The sharp break in the trend around 1973, and the corresponding increase in energy prices, provides eloquent testimony to the responsiveness of demand to price, a matter that was in some doubt in the 1970s. What remains unclear is the relation between crude oil price increases and general energy price increases. Bohi and Darmstadter do not address this point, but if the increase of other fuel prices was largely independent of crude oil prices, as I believe to be true, then an interesting corollary emerges. Just as much of the "crisis" engendered by the oil price shocks of 1973–74 and 1979–80 was caused by domestic regulatory policies, so it is that much of the subsequent reduction in energy-intensity also originated domestically.

Electricity prices illustrate the point. They did rise, more or less as crude oil prices did, but not as dramatically. At their peak in 1982–83, industrial users paid 90 percent more for electricity (in real terms) than they had ten years earlier, while residential customers paid almost 40 percent more. Electricity prices would have risen in response to the increase in crude oil prices, but higher rates would have characterized those regions that used residual fuel (e.g., the Northeast and California) and would not have caused such an increase in the national average. The increase in national average energy prices is explained by increased nuclear power costs and higher coal prices.

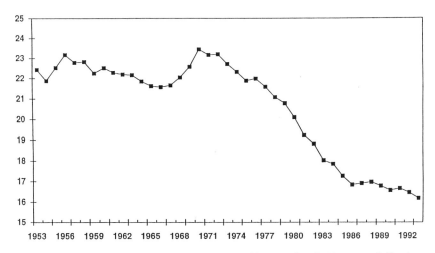

Figure 3.1   U.S. Energy/GDP Ratio, 1953–1993 (thousands of BTU/1987 dollars)
Sources: Aggregate energy data for 1953–72: *Project Independence Report, Statistical Appendix,* pp. 9–16; for 1973–93: *Monthly Energy Review,* February 1994, table 1–2.

GDP data for 1953–58: *Economic Report of the President,* January 1977, table B-10; for 1959–91: *Economic Report of the President,* February 1992, table B-2; for 1992–93: *Survey of Current Business,* February 1994, table 1.2.

Nuclear power provides the clearest example of rising energy costs that cannot be attributed to oil prices. Operating costs for nuclear power plants are higher, on average, than those for coal plants. The primary cause of this unexpected result is the cost of safety and emergency planning measures that were imposed following the accident at Three Mile Island. Demand for new nuclear plants diminished after 1973, although those on order by 1973 contributed substantially to a reduction in fossil fuel demand in the utility sector.

Coal is a tougher case. While coal prices rose precipitously in the 1970s, oil prices had little to do with the observed price changes. The relation between coal and oil prices is depicted in figure 3.2. There is a superficial similarity in that the prices of both fuels declined in the 1950s and 1960s, rose sharply in the 1970s, and have declined since. The similarity ends there. Coal prices rose sharply from 1969 through 1975, more than doubling in real terms. But roughly half of the increase occurred *prior* to 1973. By 1979–80, when crude oil prices were rising again, coal prices had already begun to fall, and the decline in coal prices continued right through the second oil price increase, despite the coal export "boom" of the early 1980s. Finally,

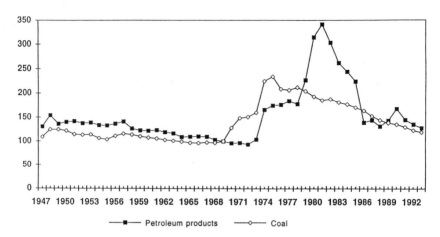

Figure 3.2   U.S. Coal and Oil Prices, 1947–1993 (nominal prices deflated, 1969 = 100)

Sources: Petroleum product prices (WPI/PPI component 057) for 1947–69: *Handbook of Labor Statistics, 1975: Reference Edition,* table 132; for 1970–82: *Handbook of Labor Statistics,* 1983, table 116; for 1983–88: *Statistical Abstract of the United States,* 1990, table 772; for 1989–92: *Producer Price Indexes,* annual editions, table 4; for 1993: estimated by author based on data for PPI component SIC 29, *Monthly Labor Review,* April 1994, table 35.

Bituminous coal prices (WPI/PPI component SIC 1211) for 1947–56: *Coal Price Formation,* December 1977, table C-4; for 1957–70: *Handbook of Labor Statistics, 1975: Reference Edition,* table 135; for 1970–78: *Handbook of Labor Statistics,* 1983, table 118; for 1979–88: *Handbook of Labor Statistics,* 1989, table 121; for 1989–92: *Producer Price Indexes,* annual editions, table 3; for 1993: estimated by author based on data for PPI component SIC 12, *Monthly Labor Review,* April 1994, table 35.

GDP deflator for 1947–58: *Economic Report of the President,* January 1977, table B-4; for 1959–88: *Economic Report of the President,* February 1992, table B-3; for 1989–93: *Survey of Current Business,* April 1993 and February 1994, table 7–1.

neither the 1986 collapse of crude oil prices nor the 1990 spurt had any effect on the trend in coal prices.

Figure 3.3 shows the primary explanation for the movement of coal prices: changes in productivity. Productivity is indicated by an index of the quantity of input, aggregated as capital, labor, energy, and materials, required to produce a unit of coal where the inputs are weighted by their share of the total value of gross output. This index of unit input is the inverse of the more usual measures of productivity, which show output per unit of input, and is used here to facilitate comparison with the price index. The index of unit input rose significantly in the 1970s, in sharp contrast to the experience in other post–World War II decades. By 1978 it required 60 percent more input to produce a ton of coal than it had nine years earlier.

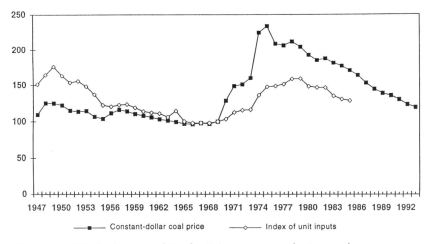

Figure 3.3   U.S. Coal Price and Productivity, 1947–1993 (1969 = 100)
Sources: Bituminous coal prices and GDP deflator: see figure 3.2. Multifactor productivity index constructed by author based on data provided by Dale Jorgenson and associates.

During the 1970s, unlike previous decades, real prices of all inputs rose, causing a further rise in coal prices.

The crude oil price increase of 1973–74 contributed to the precipitous increase in coal prices in 1974, but that year also saw a widely expected and prolonged strike by the United Mine Workers of America (UMWA), which increased the demand for inventory before the strike and restricted supply during it. Whatever the explanation for these highly perturbed years, by the late 1970s declines in productivity had come to an end, the panic effects of 1974–75 had worn off, and coal prices never again increased, notwithstanding more turmoil in crude oil markets, repeated UMWA strikes, and a significant increase in demand for coal throughout the 1980s. The causes of the coal price increase in the 1970s have not yet been adequately explained, but it should be clear that other factors were important and that they were of strictly domestic origin.

Although mine-mouth prices fell in real terms throughout the late 1970s, the delivered price of coal to electric utilities continued to rise until the early 1980s. Partly this was due to the increase in the cost of diesel fuel used to transport most coal from mine to power plant. There were other domestic causes: efforts to ensure adequate revenue for the railroads before and after deregulation in 1980, and the effect of coal supply contracts that locked in 1970s mine-mouth prices adjusted for inflation.

The evidence concerning natural gas is more mixed than that for coal,

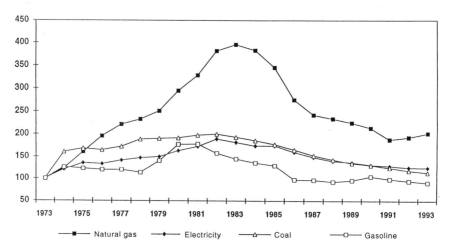

Figure 3.4   U.S. Fuel Prices, 1973–1993 (1973 = 100)
Sources: Energy prices for 1973–88: *Historical Monthly Energy Review,* 1991, tables 9–4, 9–9, 9–10, 9–11; for 1989–93: *Monthly Energy Review,* February 1994, same tables.
   GDP deflator: see figure 3.2.

nuclear power, or electricity. As the closest substitute for petroleum products, natural gas could be expected to show prices moving in tandem with oil prices. This seemed to be the case in 1986, when the prices of both fuels collapsed. Since then the two have moved quite differently. Natural gas prices did not increase notably beyond the seasonal pattern during the Persian Gulf War excursion of crude oil prices, and over the past two years natural gas prices have moved higher independently of oil prices.

   Natural gas was severely underpriced in 1973, and until recently its price was influenced more by regulatory policy than by competing fuel prices. The oil price increase of 1973–74 may have made the continued regulation of natural gas prices less viable, but it is hard to imagine that natural gas would not have been deregulated anyway. Curtailments of natural gas supplies began before 1973, and they would have continued without the oil price increase. Finally, there is little reason to suppose that the general embrace of deregulation emerging in the 1980s would have spared gas.

   Figure 3.4 shows the evolution of energy prices, as faced by industrial users in real terms, relative to the prices in 1973. Of all energy sources, natural gas is higher-priced today, and in less demand, relative to 1973 than any other competing fuel. This is due not to artificially high prices in 1993 but to artificially low prices in 1973. In addition, the most significant energy price increases relative to 1973 were registered by domestic alternatives, not gasoline or petroleum products.

The crude oil price increases of the 1970s were certainly defining events, but it is easy to exaggerate the role of crude oil and OPEC in explaining the energy crisis and the response to it. Much of the increase in energy prices observed after 1973, and much of the subsequent reduction in energy-intensity, would have occurred in any case. The OPEC price increases provided a powerful incentive to reduce demand for petroleum products, but domestic factors explain increases in the cost of nuclear power and in the price of coal, natural gas, and electricity.

The Contribution of Research and Development

R&D is one aspect of the policy response to the energy upheavals of the 1970s that Bohi and Darmstadter pass over too lightly. One of the early policy responses to the energy "crisis" was the creation of the Energy Research and Development Administration (ERDA) with its mission to foster technology to reduce the nation's dependence on imported oil and, more grandiosely, to effect the transition away from depletable energy resources. ERDA's mission was transferred to the Department of Energy (DOE) in 1977 and accounted for a major part of that department's budget. Unlike most of the other policy responses to the energy upheavals of the 1970s, energy R&D is still in place and actively funded, albeit at a reduced level.

While Bohi and Darmstadter discuss synthetic fuels, they do so in the context of the Synthetic Fuels Corporation rather than discuss the funding of the same technologies within ERDA/DOE. There was much more to energy R&D than synthetic fuels. Breeder technology was a major focus of R&D activity for many years, and there was significant funding for renewables, conservation, and alternative transportation technologies. The reasons for federal funding of R&D are good ones; and given the significance of technology, measurable accomplishments could reasonably be expected. The views of informed and judicious observers such as Bohi and Darmstadter on this aspect of energy policy would have been appreciated.

The authors' neglect of the subject is revealing. After twenty years nothing impressive emerges from the energy R&D effort. In the main, the energy technology of the 1990s is an improved version of what existed in 1973. The only notably different technology is the gas-fired combined-cycle power plant. Although there was some ERDA/DOE funding of alternative bottom and topping cycles, the technology used today originated in Department of Defense R&D after a considerable gestation period in the aerospace industry.

The Efficacy of Government

In the last twenty years society's belief in the efficacy of government has changed. The balance between public and private sectors has shifted significantly to the point where traditional instruments, such as taxes, access to public lands, and the regulation of health and safety, are virtually the only means by which government can directly affect the energy industries. Price and allocation controls for petroleum products are gone, as are mandates for using certain fuels. Federal regulation of wellhead prices and interstate transmission tariffs for natural gas are effectively dismantled. And in 1994 the dominant energy policy issue is deregulation of the electric utility sector. Markets are being created to achieve environmental goals where previously no thought was given to anything but traditional command and control regulation.

The response to the energy upheavals of the 1970s was in many ways the last great experiment in public policy. It was a time when the citizenry and the political elites seemed to believe that government action could ameliorate untoward circumstances created by events in the Middle East. No doubt today's disenchantment with government is part of a broader shift in attitude that has been building over the past two decades. The nation's experience with energy policy did not by itself cause this change. Still, energy policy provided unforgettable examples that crystallized latent sentiment and perhaps hastened the process.

Two particular aspects of energy policy fostered disenchantment with government. The first is the signal failure of what was presented as energy policy. Small and perhaps redeeming successes can be found, but in the main, performance fell so far short of promise that it is hard to view Richard Nixon's Project Independence or Jimmy Carter's "moral equivalent of war" with anything but amusement. The policy successes, ironically, were those that reversed earlier government actions—notably, deregulating the wellhead price of natural gas, ending oil price and allocation controls, and chipping away at the electric utility franchise.

Policy proposals that extended the reach of government were for the most part embarrassing failures. The infatuation with synfuels, culminating in the Synfuels Corporation, would head any list, but there are many close seconds: the Powerplant and Industrial Fuel Use Act; entitlements; the Strategic Petroleum Reserve, whose failure has yet to impress itself fully upon us; and the significant expansion of R&D funding for energy technology. Efficiency standards, whether applied to cars, buildings, or appliances, may have been more successful. They have not been embarrassing failures, and they did extend the reach of government, but there is a hint of superfluity about them—namely, as was argued when proposed, they were needed because energy prices were controlled.

A second way in which energy policy has contributed to a loss of faith in the efficacy of government is the spectacle of the private uses of government action. Because energy is a pervasive element of modern society, the energy policies of the past accommodated many agendas. It was not long before energy policy lost any consistent cognitive content, as it was called upon by proponents of diverse identity to justify any plausibly related proposal. The promotion of ethanol is the most successful example of this phenomenon, but there have been many other attempts, not the least of which is the use of energy policy as justification for various energy taxes to reduce the budget deficit.

In conclusion, as Bohi and Darmstadter note, advocates still seize upon energy policy to advance particular interests. An encouraging trend today, however, is that these proposals meet with little credence from the disinterested policy community and citizens at large.

REFERENCES

Coal Price Formation. 1977. Palo Alto, Calif.: Electric Power Research Institute, December.

Economic Report of the President. Various years. Washington, D.C.: U.S. Government Printing Office.

Handbook of Labor Statistics. Various years. Washington, D.C.: U.S. Department of Labor, Bureau of Labor Statistics.

Historical Monthly Energy Review, 1973–1988. 1991. Washington, D.C.: U.S. Department of Energy, Energy Information Administration, Office of Energy Markets and End Use.

Monthly Energy Review. Various issues. Washington, D.C.: U.S. Department of Energy, Energy Information Administration.

Monthly Labor Review. Various issues. Washington, D.C.: U.S. Department of Labor, Bureau of Labor Statistics.

Producer Price Indexes. Various years. Washington, D.C.: U.S. Department of Labor, Bureau of Labor Statistics.

Project Independence Report, Statistical Appendix. 1974. Washington, D.C.: U.S. Federal Energy Administration.

Statistical Abstract of the United States. 1990. Washington, D.C.: U.S. Government Printing Office.

Survey of Current Business. Various issues. Washington, D.C.: U.S. Department of Commerce, Bureau of Economic Analysis.

Verleger, Philip K., Jr. 1993. *Adjusting to Volatile Energy Prices.* Washington, D.C.: Institute for International Economics.

4. Watershed, Aberration, and Hallucination

The Last Twenty Years

ALLAN PULSIPHER

As Doug Bohi and Joel Darmstadter insightfully point out in chapter 2, most aspects of the energy upheavals of the 1970s and 1980s can best be described as both an aberration and a watershed. I would add that there were also aspects that should be described as hallucinations. I understand *watershed* to represent a clear change in observable, explainable trends, and an *aberration* to be a temporary departure from these trends. I use *hallucination* in the classic sense of a (usually bizarre) distortion of, and loss of contact with, reality. I add *hallucination* because I believe one aspect in particular—hard and soft energy—continues to confuse some who are concerned with energy and environmental issues.

The Most Significant Watershed

President Ronald Reagan's campaign pledge to do away with the Department of Energy and, when Congress balked, to appoint a dentist as secretary of energy[1] was the most significant watershed in modern U.S. energy policy. Whether it was a consequence of insightful strategic analysis or of blind ideology, this watershed clearly and *explicitly* signaled that federal finagling with energy prices, supplies, and the nation's fuel mix—which had provided the strategic continuity for national energy policy throughout the Nixon, Ford, and Carter administrations—had come to an end. *Implicitly* this policy watershed also meant that President Jimmy Carter's "moral equivalent of war" had been restored by President Reagan to the "military equivalent of war, if necessary."

Eventually, Iran's ayatollah proved that even the purest of heart and most vengeful of spirit really had no use for crude oil other than to sell it to the infidels for what it would bring in the market. Thus we came to understand that, operationally, "if necessary" actually meant "if needed to keep the vast amounts of money to be made from the sale of oil out of the pockets of those who would use it to do us harm." As in most other realms, it was money or something else, not oil, that was ultimately at issue.

Fourteen years later this simple "market-cum-military" conjunction still defines U.S. energy policy. Although rhetoric changed when the Democrats regained control of the presidency in 1992, the fundamental reality remained. U.S. energy policy, as concisely described by Louisiana senator J. Bennett Johnston, chairman of the Energy and Natural Resources Committee, has been distilled to two words: import oil.

This concise description is commonly interpreted as criticism. However, such an interpretation misses its deeper economic implication. Why do we import so much oil? Because it is so cheap. *Not* to import so much oil would be more expensive. Given a choice, we as a nation prefer to spend as little of our money as necessary to acquire the energy we need. This has most recently been affirmed during the thoughtful public consideration of the Clinton administration's proposed "BTU tax."

I do not find the usual criticism of the market-cum-military policy—that it reflects only the cheaper imported energy prices and ignores the costs of military readiness and intervention—very persuasive. Although the military costs implicit in implementing the policy are very real, and military intervention raises relevant ethical and political issues, I am not sure that there are *additional* military, or "ethical," costs attributable to the strategy.

Would the Department of Defense's budget be much different if the Department of Energy were still in the synfuels business, or if Congress had adopted 50-mile-per-gallon CAFE standards, or if the United States imported only 25 percent rather than 50 percent of its crude oil? Does the cost of maintaining the ability to intervene militarily in Somalia, Bosnia, Mexico, or Korea differ significantly from what it would have been if we did not also want to keep our multibillion-dollar oil import payments out of the pockets of Iraq's Saddam Hussein? I have not seen a persuasive case made for the affirmative.

Looking back at the efforts to think through the energy problem in the late 1970s, I am struck by the absence of consideration of the now-reigning market-cum-military strategy. Despite the fact that it was adopted within two years' time by Reagan and retained by subsequent administrations— along with the Democratically controlled Congress—I could not find it in the troika of energy books that appeared almost simultaneously in 1979:

Energy in America's Future, by Sam Schurr and others; *Energy Future,* by Robert Stobaugh and Daniel Yergin; and *Energy: The Next Twenty Years,* by Hans Landsberg and others (Schurr et al., 1979; Stobaugh and Yergin, 1979; Landsberg et al., 1979).

I confess that my "method" was to look up *military* in the indexes of these books, but the word apparently was not used in either the Schurr or the Landsberg volume, and the only reference in Stobaugh and Yergin was to the Military Construction Act of 1979, which said that all military housing and 25 percent of all other military construction was to incorporate solar heating if economically feasible.

Both Schurr and Landsberg discuss at considerable length the virtues of letting the market allocate energy resources and the various dimensions of what we now refer to as energy security. But neither evaluates the simplest, purest extension of the argument: leaving allocation and distribution of energy supplies to markets and letting the State and, if necessary, Defense Departments worry about energy security.

I am not sure whether those who participated in these efforts failed to see the impending relevance of such a heavy-handed strategy or were too intimidated by prevailing conventional wisdom to take it seriously. How prudent this watershed decision will prove for our children is a question I leave for the future meetings of the Council on Foreign Relations; but to my mind, the decision to leave energy to the market—and, if necessary, the military—is clearly the key watershed of the past two decades.

The Most Significant Aberration

Whether the path of oil prices should be considered a watershed or an aberration depends on how one views the Seven Sisters' semi-incestuous relationship with the Texas Railroad Commissioners. From the close of World War II until the Arab oil embargo of 1973 (see fig. 2.1), this relationship kept the price of oil on a track that looked like the electrocardiogram of a room-temperature patient. Focusing only on the nominal market price of oil on an average annual basis, the stability of the two decades preceding the energy shocks is more remarkable. The nominal price normally changes by about a penny a barrel from year to year.

If this comalike stability is viewed as evidence of a managed and manipulated but still market-determined or market-responsive price, then one comes down on the side of aberration. Oil price movements from 1978 to the present have become progressively easier to understand by tracking traditional market factors influencing oil supply or demand rather than by trying to unsnarl longer-range political or ideological motivations or theories about the strategic dynamics of cartels.

The energy upheavals of the 1970s taught us, eventually, that the lessons we learned in Economics 101 in the 1960s were true. Cartels are inherently unstable and ultimately futile. Energy markets are more competitive today than they have been since the end of World War II. Assumptions much more consistent with competition than with cartels drive today's energy price forecasts along, or below, the lower limit of the range forecasters in the 1970s used to reflect the uncertainty in the international oil market.

Among the more interesting intellectual artifacts from the energy policy frenzy of the 1970s is William Nordhaus's analysis of the optimal allocation of energy resources across time, beginning in 1973 with his report in the *Brookings Papers on Economic Activity* (Nordhaus, 1973: 529–76) and further elaborated in a Cowles Foundation monograph, *The Efficient Use of Energy Resources* (Nordhaus, 1979).

Nordhaus compared several oil price forecasts that tried to build on estimates of varying degrees of market power, including two of his own. He also provided an efficient or competitive forecast—more as a reference point than an actual forecast. He estimated that, measured in 1975 dollars, the price of oil in a competitive market in 1995 would be \$7.02/barrel. Using the most recent producer price index for crude energy materials to translate Nordhaus's competitive "forecast" into 1993 dollars yields a \$14.53/barrel "estimate" for 1995. I expect this to be only a little below the mean value of the next summary of short-term forecasts appearing in publications such as *Petroleum Intelligence Weekly* or *Oil and Gas Journal*.

Nordhaus's two monopolistic forecasts, updated to 1993 dollars, were \$30.53 and \$37.66/barrel, well beyond the upper limit of the aspirations of even the most optimistic, paranoid, or xenophobic of today's producers, foreign or domestic. Current estimates of a competitive, market-determined world oil price fall in the \$10/barrel range. However, Nordhaus's twenty-year-old competitive projection looks like it fits the facts much better than his (or anyone else's) cartel-based predictions.

Nordhaus's analysis contained another inadvertently prescient observation. Almost apologetically, he wrote, "It is somewhat surprising . . . that nuclear generation of electricity is delayed [in his optimal or efficient world] until 2000." This tardiness, he explained, resulted from much higher petroleum prices in the real world than in his efficient solution, from nuclear subsidies, and from "the rapid and unexpected run-up in prices of nuclear generating equipment reflected in the technological assumptions" (Nordhaus, 1973: 553). Hendrik Houthakker, a member of President Richard Nixon's Council of Economic Advisers, stated, "I agree with Nordhaus' timetable, according to which nuclear energy is not relevant until the turn of the century. The nation developed nuclear energy prematurely. . . . I doubt that many nuclear power plants would have been built in a completely free market" (Nordhaus's discussion, 1973: 572).

Gifted economists often know more than they think they know, and the rest of us, including those hooked up to the Tennessee Valley Authority's power system, could save a great deal of money if there was some way of knowing when this knowing occurs.

The Most Significant Hallucination

My nominee for the most significant hallucination in energy policy is a body of thought that for want of a more elegant term I call furnace marxism. Like real marxism, furnace marxism views social and political institutions and attitudes as dancing to a deeper dynamic ultimately driven by technology that is connected and controlled by prevailing economic relationships and endowments. Furnace marxists, however, are more discriminating than classical marxists. It is primarily the energy economy, not the much larger general economy, that is the prime mover.

There are "hard" and "soft" furnace marxists. Lyndon LaRouche's obsession with fusion energy may be the best example of hard-energy furnace marxism, but spokespersons from the energy industries also often attribute greater-than-simply-economic consequences to the origin, cost, or availability of particular energy sources, in terms that have discernible overtones of furnace marxism.

The most influential furnace marxists are soft-energy enthusiasts. For them, large nuclear or fossil-powered electricity-generating facilities, petroleum refineries, dams, and regional electricity grids all lead to authoritarian societies that threaten individual freedom and the integrity of ecosystems. Smaller, dispersed energy systems preserve liberty, promote self-reliance and democratic institutions, and protect ecological balance. Although rarely articulated, the soft-energy vision has gained almost the status of a core ideological assumption in a large spectrum of contemporary social criticism.

I can personally relate to soft furnace marxism. Early in my life as a scholar, economic circumstances required that I live in a house warmed by a coal furnace. The coal came from a mine located less than three miles from my house and was dug and delivered by my neighbors. In time the daily satisfaction of the stoker's appetite for fresh coal and the wrestling of the clinker from the furnace's firebox became familiar stress-reducing rituals. During the heating season my family and I were frequently exposed to air that violated primary air-quality standards. But the impersonal, grid-dependent gas or electric furnaces that I have subsequently had to depend on for warmth have never provided the same reassuring independence I got from my day-to-day relationship with that coal furnace. Setting this personal observation aside, the analytical and empirical support for hard or soft furnace marxism, in my opinion, ranks a notch or two below that for chiropractic medicine or astrology.

One of the interesting peculiarities of both soft and hard furnace marxists is the asymmetry of their optimism about energy technology. Those of the soft-energy persuasion find not only that great progress had been made in technologies that make decentralized energy use and energy conservation more feasible, but also that the technologies on the drawing boards look even better. However, similar technical progress and optimism never seem to apply to larger, harder, centralized energy technologies. A fuel-cell-powered automobile, one that is cheaper than a Yugo but that drives like a Lexus, is just around the corner. But the prospects for cheaper, more dependable nuclear reactors simply seem to get worse.

The other dimension of this asymmetry is the notion that technologies keep being developed that would solve most if not all of our energy and environmental problems if private or governmental conspiracies did not keep them from us, or if we simply were not too dumb to adopt them. This attitude is epitomized in the National Academy of Science's finding that "the U.S could reduce greenhouse gas emissions by 10 to 40 percent of the 1990 level at very low cost. Some reductions may even be at a net savings" (National Research Council, 1991: 63). Graphics that serve as the basis for this statement purportedly show that, at the more optimistic limit, the nation could save money by reducing greenhouse gas emissions up to 25 percent of the 1990 total. Given the prudence of this goal, one wonders how a society clever enough to develop such technology can be so collectively inept or retarded that it permits, at considerable cost to itself, its contribution to global greenhouse gas concentrations to increase.

The environmental or economic values and judgments that soft- (or hard-) energy furnace marxists share are not necessarily incorrect or without merit or empirical support. Like most markets, energy markets are subject to imperfections and biases. These imperfections can and should be identified and, if warranted, corrected on a case-by-case basis. But there are neither theoretical nor empirical arguments that particular mixes of energy sources and uses, or particular degrees of energy efficiency or inefficiency, are important or fundamental determinants of the type of society or economy a country has. Available evidence from careful international energy comparisons of the sort pioneered by Joel Darmstadter and his colleagues at Resources for the Future simply does not support such claims. Japan and France made the largest relative contributions toward reducing greenhouse gases in the 1980s—not via soft energy, but by significant increases in nuclear power for domestic electricity.

Conclusions

Energy matters provide an interesting collection of policy problems and questions that are largely economic and technical in nature and that vary with particular circumstances. All-encompassing pseudo-ideologies such as soft or hard energy rarely provide an adequate framework for asking the right questions or finding efficient and equitable answers to them. Unfortunately, the real world is more complex than these one-dimensional slogans suggest.

Energy policy issues probably reached their peak popular currency in the late 1970s. After President Carter tried to elevate them to the level of a moral crusade, speculation and panic regularly drove world oil prices to new highs, and wooden domestic allocation schemes made waiting in line to buy gasoline a too-familiar part of the urban experience. At that time I made my living as a program officer in the Energy and Environment Division of the Ford Foundation, helping to finance energy and environmental studies, symposia, seminars, and workshops. A question that regularly surfaced at such gatherings was, Where are the disadvantaged, the poor and the blacks (or their representatives)? Don't they realize the importance of energy? As a Jeffersonian Democrat, I was driven to the conclusion that they must have more important things to worry about. My "answer" was usually treated as a joke rather than as simple heresy. But I was right. They still do. We should not lose sight of that reality twenty years later.

NOTE

1. This description is facetious. The individual referred to was experienced, politically active, and knowledgeable about energy prior to his appointment. However, his visibility and his perceived clout within the energy cognoscenti were easily an order of magnitude below those of James Schlesinger, President Carter's secretary of energy.

REFERENCES

Landsberg, Hans, et al. 1979. *Energy: The Next Twenty Years.* Cambridge, Mass.: Ballinger.
National Research Council. 1991. *Policy Implications of Greenhouse Warming.* Washington, D.C.: National Academy Press.
Nordhaus, William. 1973. *The Allocation of Energy Resources.* Brookings Papers on Economic Activity no. 3. Washington, D.C.: Brookings Institution.
———. 1979. *The Efficient Use of Energy Resources.* New Haven: Cowles Foundation, Yale University Press.
Schurr, Sam H., et al. 1979. *Energy in America's Future: The Choices before Us.* Baltimore: Johns Hopkins University Press, for Resources for the Future.
Stobaugh, Robert, and Daniel Yergin, eds. 1979. *Energy Future: A Report of the Energy Project at the Harvard Business School.* New York: Random House.

5. Reflections on the Twentieth Anniversary of the 1973–1974 Oil Price Shock

GEORGE HORWICH

We are constantly reminded that success has many parents; failure is an orphan. This is certainly true of oil and gas deregulation in the United States today—a success by almost any criterion. Virtually everyone, most notably consumers, embraces deregulation's aftermath: continuing weakness in world crude oil prices, the constant decline in the inflation-adjusted prices of petroleum products, restoration of an efficient and reliable fuel distribution system, and the huge, primarily market-driven reduction in the energy/GDP ratio. Moreover, as Douglas Bohi and Joel Darmstadter note in chapter 2, this embrace of deregulation is particularly gratifying because it extends to politicians and other government officials who formerly favored more regulation during the mid-1970s. Now these groups appear to have genuinely, if reluctantly, absorbed one of the lessons of the energy crisis noted by Bohi and Darmstadter: that government's ability to alter economic behavior is limited. Yet in recalling the energy crisis of 1973–74, it should be remembered that not all the support for energy regulation came from politicians and unsophisticated government officials. Many economists and other policy professionals, including some in high government and academic posts, supported—in part, if not totally—price controls and mandatory allocations. They lent, at most, lukewarm support to the decontrol policies of the late 1970s. Their reasons for wanting to retain controls were sometimes quite sophisticated.

President Jimmy Carter's Council of Economic Advisers (CEA) feared that in the absence of controls, an oil disruption would raise prices and unleash massive consumer spending on oil products. The resulting revenues would be logjammed in oil producers' coffers, severely reducing aggregate demand in the economy at large. This phenomenon, known as "oil price drag," was a concern expressed by CEA reports beginning in 1974. Thus, while the January 1980 report endorsed the president's gradual decontrol initiative that began in June 1979, that approval was qualified (*Economic Report of the President*, 1980: 107–9). While recognizing that decontrol would lead to more rational prices, the report urged the establishment of a standby fuel-rationing plan and, by implication, price ceilings on gasoline, diesel fuel, and heating oil for use in a major disruption (p. 106).

The extent to which oil price drag reinforces the negative GDP effects of an oil price shock can only be determined empirically. In spite of its deep and continuing concern with the possibility of drag, the 1980 CEA report offered only back-of-the-envelope estimates of the GDP impact of drag (*Economic Report of the President*, 1980: 65). In 1981 and 1982 I carried out a full-scale measurement of drag, using oil-industry balance-sheet and income data furnished by the American Petroleum Institute and the Energy Information Administration. Year-by-year estimates for 1969–80 ranged from one-sixth of CEA's estimates to, at most, 62 percent (Horwich and Weimer, 1984: chap. 6). In general, oil companies, in both their domestic and foreign operations, proved remarkably able to spend and return receipts, of any magnitude, to domestic goods markets without delay.

Even without detailed knowledge of the likely size of oil price drag, it remains a mystery why Carter's CEA did not promote a traditional offsetting adjustment, such as an easier monetary policy, to counteract an essentially macroeconomic disturbance—the possible weakness in aggregate demand due to increased holding of money by the oil industry. A complex network of price controls and rationing, with all its intricacies of implementation, was clearly a less efficient policy alternative.

Less defensible than its preoccupation with oil price drag was the CEA's complaint that oil price decontrol would create producer windfalls that would be ineffective in increasing supplies and that would reduce real consumer incomes (*Economic Report of the President*, 1980: 109). These assertions are utterly at odds with conventional economic analysis, which assumes positive price-elastic supply curves and an increase in social surplus whenever price ceilings are lifted.

The CEA's desire to impose controls on petroleum and petroleum products in the event of another disruption was based only partly on its fear of oil price drag. A concern, based on an opposite tendency in the price level and shared with many Keynesian economists, was that inflation sparked by oil price increases would create widespread inflationary expectations. Once reflected in the economy's contracts, these expectations would almost certainly mean that future attempts to reduce inflation would lead to costly recessions (Tobin, 1979). The Keynesians did not recommend long-term use of ceilings, recognizing that distortions would gradually build up (though they believed that the distortions could be minimized with a good rationing scheme). But they felt that a short period of controls, even with distortions, was worth trading for future control of the price level.

Monetarists opposed controls on principle, but not many showed a very deep understanding of what a supply-side shock, such as an oil disruption, was all about. They frequently argued that if the money supply were properly controlled, any increase in oil prices would be offset by decreases in other prices (see Horwich, 1991: 143–44). That can only occur if the rise in oil prices is purely a demand phenomenon, resulting from a shift of demand to oil, whose price rises, and away from nonoil products, whose prices are downwardly flexible and have fallen. On net, both aggregate demand and aggregate supply are fixed, and the general price level is unchanged. A supply shock, on the other hand, causes a leftward shift of the aggregate supply-of-output schedule along a fixed aggregate demand schedule. Both the relative price of energy and the absolute general price level are higher as a direct result of the disturbance.

Joe Kalt showed that the flaw in the Keynesian position was that controls, particularly as designed in the 1974–81 period, raised oil imports by about a third and caused a sizable increase in the world price of crude oil and petroleum products. Controls raised imports by discouraging domestic production and raising total oil demand by way of entitlements. Entitlements equalized prices of domestic and imported crude oil, artificially reducing the price on imported oil faced by domestic purchasers (Kalt, 1981: 191, 199–201). But even without entitlements, crude oil controls would have diverted excess domestic demand into the foreign market, raising world prices.

Controls and Equity

Economists generally avoided endorsement of the Byzantine allocations and monetary transfers mandated under the controls. Many policy professionals, however, including some at our best think tanks and universities, saw merit in them. One transfer often discussed and sympathetically regarded was the huge network of subsidies given to small, at the expense of larger, refiners. The justification was that the small operators were less successful in bidding for and obtaining the more limited supplies of crude oil following disruptions. During the 1970s and 1980s I recall no awareness in academic or policy circles that "access" to crude oil was related to demand for refiners' products and that small refiners faced more elastic demand schedules than did large ones. Small refiners generally produced at the "heavier" end of petroleum products (e.g., lower-grade heating oil, boiler fuel) for which there were some shorter-term substitutes, such as coal and natural gas, when prices rose. The output of larger refiners tended to be concentrated at the lighter end (e.g., jet fuel, gasoline) for which direct substitutes were far fewer and demand less elastic. In disruptions, product prices thus tended to rise less and demand quantities to fall more for the output of small than for that of large refiners. This left the small ones with weaker incentives to bid aggressively in the disrupted crude oil market, to stockpile crude oil, or to secure crude oil supplies through predisruption contractual arrangements. In the opinion of many, the lesser access of small refiners was a matter of smaller size and more limited resources—an equity issue, which mandatory allocations properly addressed. Differing access was not seen as a rational response to differing product demand elasticities, and therefore was not seen as an obvious public policy issue.

Some economists and policy analysts, while recognizing that price controls created serious inefficiencies, nevertheless believed that controls generated compensating benefits to lower-income families. Studies of the equity impact of the 1970s regulations concluded that the well-to-do and the well-connected were likely to receive most of the benefits obtainable under non-price-controlled allocation (Landsberg et al., 1979: 18–26; Horwich and Weimer, 1984: 101–4). The well-off provided their teenage offspring with multiple automobiles to fill gasoline lines; persuaded Congress that even they could be "small" refiners deserving of continued support; and, as established distributors, received product allocations based on historical use patterns, preventing new competitors from gaining a foothold. The poor, when finally able to reach the pump, did not pay a price significantly lower than the uncontrolled free-market price because of the tendency of controls to raise the world price of crude oil. Moreover, the poor and less skilled suffered the brunt of job loss due to inefficiencies caused by the controls and affecting the economy generally.

Blood for Oil?

The suggestion by some that future energy policy is likely to show more respect for uncontrolled markets in disruptions but to include a military option is not well founded, at least in the latter assertion. We did indeed allow prices to clear oil markets following the Iraq-Kuwait flareup in August 1990. There was the usual anti-industry jawboning from the White House, but contrary to at least one analysis (Verleger, 1990), there is no evidence that transactions were carried out under other than effectively free-market conditions (Horwich, 1992: 28–31). However, although the country quickly mobilized and responded to the invasion of Kuwait, security of the oil fields for use by the United States and its allies was not the real military objective (pp. 26–28). It is true that U.S. political leaders stated that protection of the fields was one of the goals, and the intervention may have accomplished that. But this outcome was an incidental by-product of another mission. The real objective of our intervention was not to secure the oil for ourselves or other consuming countries; rather, it was to deny the benefits of its ownership to Saddam Hussein.

Between August and October 1990 the price of crude oil rose 118 percent on world markets. This was a more precipitous rise than had occurred even in 1979–80. Nevertheless, it is unconvincing to argue that this slightly more than doubling of the price merited a military response while the twelvefold nominal increase that occurred from 1973 to 1981 did not. During the oil crises of the 1970s there were no serious moves in Congress or the White House to mobilize troops. We viewed OPEC's control of the world market and the market's later destabilization by the Iranian Revolution and the Iran-Iraq War as economic disturbances for which only diplomatic and economic responses were appropriate. Only verbal missiles were fired in those engagements.

Why was 1990 different? Although the Iraqi invasion and subsequent United Nations embargo removed 4 million barrels (about 8 percent) of oil from the noncommunist world supply, the real price of crude oil (corrected for inflation) had fallen throughout the 1980s to roughly double its 1973 predisruption level. More important, the real prices of petroleum products, such as gasoline and heating oil, were back to pre-1973 levels. OPEC's share of world production, moreover, was only 38 percent, compared with 56 percent in 1973. Nothing in these numbers suggests that consumer countries faced an economic burden greater than that of the 1970s. Nor was Iraqi-Kuwaiti output permanently lost to the world even if controlled by Saddam.

Even tyrants avidly seek revenues, and there was every expectation that Saddam would produce the fields under his control, including those in Kuwait and, potentially, Saudi Arabia. He would surely have restricted output below previous levels, but it was unlikely that his optimum price exceeded $32/barrel—hardly a cause for war (Henderson, 1990). One might question the stability of his production, given his predilection for military force, but there were ample oil reserves in the consuming world to offset any instability.

Willing to seize the fields of his neighbors rather than negotiate friendly cartel agreements with them, Saddam was perceived as a lethal threat to the Middle East and to all populations within reach of the missiles he either had or was building. The oft-repeated suggestion that we would not have moved militarily to protect Saudi Arabia and free Kuwait if they produced soybeans instead of oil misses the point. Saddam would not have seized a soybean kingdom unless soybeans, like oil, commanded a rent on world markets sufficient to fuel his war machine.

Thus, while we might anticipate free-market responses for shorter-run interruptions, and if we are willing to draw our petroleum reserve for longer periods, we are unlikely to see the circumstances that trigger a military reaction by the oil-consuming countries anytime soon.

Conclusions: The Strategic Petroleum Reserve Politicized

One final lesson of the Kuwaiti war is the apparent politicization of the Strategic Petroleum Reserve (SPR) (Horwich, 1992: 32–35). With 600 million barrels in our SPR and a like amount in the reserves of other OECD countries, the runup in oil prices in 1990 could have been easily and quickly reversed. The usual excuse for inaction is that the administration feared a prolonged crisis, still higher prices, freeloading by other consuming countries, and so on. We did not hear much of this. The pronouncement in fact made by the White House that it did not want to exercise "control" over petroleum prices by drawing on the reserve must have been embarrassing to its economists.

Unable to obtain better information than what was published in the print media, I offer as pure conjecture an alternative explanation for SPR nonuse: the administration did not *want* to bring the price of oil down in the fall of 1990. It is easy to forget how beleaguered George Bush was in his crusade against Saddam. Bush seemed unlikely to get either congressional or United Nations support for military intervention. In that context a suitably high price of oil served as a mighty beacon reminding the world what a menace Saddam was to its peace and security.

At the same time, Saudi Arabia in the early fall committed itself to reimbursing a portion of U.S. military expenditures. The high price of oil was thus, in effect, *our* high price. Indeed, the $17 billion in additional revenues taken in by the Saudis because of the price runup were fully remitted to us after the war as compensation for our involvement. *Au revoir,* SPR.

NOTE

I thank Milt Russell and Dave Weimer for helpful comments on this chapter.

REFERENCES

Economic Report of the President. 1980. Washington, D.C.: U.S. Government Printing Office, January.

Henderson, David R. 1990. *Do We Need to Go to War for Oil?* Foreign Policy Briefing 4. Washington, D.C.: Cato Institute, October 24.

Horwich, George. 1991. "Macroeconomics and Macroeconomists as Instruments of Policy." Pp. 127–57 in *Policy Analysis and Economics,* edited by David L. Weimer. Boston: Kluwer Academic Publishers.

——. 1992. "Energy Policy, Oil Markets, and the Middle East War: Did We Learn the Lessons of the 1970s?" Pp. 25–39 in *International Issues in Energy Policy, Development, and Economics,* edited by James P. Dorian and Fereidun Fesharaki. Boulder: Westview Press.

Horwich, George, and David L. Weimer. 1984. *Oil Price Shocks, Market Response, and Contingency Planning.* Washington, D.C.: American Enterprise Institute.

Kalt, Joseph P. 1981. *The Economics and Politics of Oil Price Regulation.* Cambridge: MIT Press.

Landsberg, Hans., et al. 1979. *Energy: The Next Twenty Years.* Cambridge, Mass.: Ballinger.

Tobin, James. 1979. "Why the Fed's Cure Won't Work." *New York Times,* November 11, p. F18.

Verleger, Philip K., Jr. 1990. "Understanding the 1990 Oil Crisis." *Energy Journal* 11 (4): 15–33.

6. How Much Damage Did the Oil Price Shocks Do?

JOHN M. BERRY

Two decades after the Arab oil embargo, it is easy to forget the shock inflicted on the American psyche: the soaring prices and a sudden fear that the nation's quality of life was in jeopardy. Some of that fear was justified because the United States stood on the threshold of nearly twenty years of economic disorder marked by the highest inflation in a century, a virtual disappearance of productivity growth, and sweeping changes in major elements of the American economy, including its financial system. Much of this disorder was due to the energy price shocks.

The public, media, and policymakers shared a woeful lack of understanding about the energy shocks. One need only recall the televised shots of tankers on the Delaware River waiting to unload cargoes, cited as evidence that there was plenty of oil but that oil companies were keeping supplies off the market as part of a conspiracy to force up prices. When I went to the *Washington Post* at the beginning of 1979, many staff reporters believed in the conspiracy. Few journalists, much less the public at large, had any notion in the 1970s of how commodity markets routinely behave in the face of uncertainty about future supply. Few politicians were prepared to let the market hold sway.

Discussion

I agree with the conclusion Doug Bohi and Joel Darmstadter reach in chapter 2: the upheavals were both a departure from and a part of already established trends. However, I believe they underestimate the ultimate damage the energy price shocks inflicted upon the U.S. economy.

The price shocks, poor policy responses to them, and a significant amount of bad luck all contributed to a persistent rise in inflation that led to three recessions in eight years. Equally important, the inflation and developments related to the recycling of financial surpluses that arose in the oil-exporting nations severely threatened the viability of the U.S. financial system. Moreover, the price shocks played a significant though indirect role in getting the federal government's fiscal policies wildly off track in the 1980s and opened the door to a massive wave of new foreign competition for American firms.

The energy price shocks alone did not cause this havoc. Poor policy was also to blame. However, the abrupt, unexpected challenge presented by the energy upheaval made the execution of good policy exceptionally difficult, particularly in a democracy with highly decentralized wage- and price-setting traditions. There were no really good policy options.

The first problem was that energy price increases were similar in their macroeconomic impact to a tax increase: they reduced real incomes and restrained economic activity. The natural impulse of policymakers was to take steps to head off recession. At the same time that higher energy prices pushed the economies of the United States and many other nations into recession, the same price surge significantly increased the inflation rate. Policies dealing with one problem worsened the other, at least in the short run.

The public generally did not understand what was happening, nor did most politicians. Some policymakers had a better grasp of the poor choices confronting the nation, but even those who tried were unable to convince elected and appointed officials that the best option was to let higher energy prices be fully reflected in the costs paid by energy users. Everyone wanted protection against real income losses. The arguments presented ranged from fairness and compassion to economic efficiency and greater long-run energy security.

As is often the case, an unwillingness to make hard choices—and the politics of the moment precluded making such choices—increased the size of the nation's loss. In this case, poor policy response led to higher inflation and delayed adjustment of energy production and consumption rates. Without the energy shocks, policymakers at the time likely would not have had to deal with problems of such magnitude. As a result, the nation's economic course would have been less trying.

As Bohi and Darmstadter acknowledge, the world's energy problems did not strike out of the blue with the oil embargo of late 1973. In 1972, Treasury Secretary George Shultz openly worried that the world would be short of oil unless a way was found to enable Saudi Arabia to increase production to at least 20 million barrels a day from a level less than half that amount. Oil prices were beginning to rise, and the Nixon administration tried to find a way to keep prices stable while avoiding any threat to continued economic growth. There was no talk, at least publicly, of letting prices rise to encourage more production or to reduce demand. In addition, the 1950s and 1960s saw exceptional oil price stability, which historically had varied substantially from year to year (Verleger, 1993: 24).

In 1972, a presidential election year, a broad program of wage and price controls was in effect, interest rates were low, and the economy was booming. President Richard Nixon imposed wage and price controls in August 1971 when inflation was less than 5 percent. His authority to do so came from a law that originated with congressional Democrats. The latter firmly believed that the economy could be operated at a higher level of resource utilization without inflation getting out of hand if an incomes policy— either some type of formalized persuasion or outright controls—was used. Bohi and Darmstadter state that wage and price controls imposed to deal with inflationary expectations after the 1970–71 recession did not significantly reduce inflation. The problem, however, was that the recession was very mild.

To this day, one of Nixon's principal economic advisers, Herbert Stein, maintains that the National Bureau of Economic Research should not have counted the episode as a recession and would not have except for the depressing effect of a strike against General Motors. But in the summer of 1971 Nixon was being told that inflation was likely to worsen, particularly if action were taken to stimulate the economy. Since the president wanted to go to the voters in 1972 with a strongly growing economy, controls appeared to provide a way to deal with inflation at the same time fiscal policy moved to stimulate economic activity.

The timing of controls was apparently dictated by the country's worsening international payments position. The United States was still trying to maintain a value for gold of $35 an ounce as the foundation of the world's system of currencies. However, with U.S. inflation rising, many people in other countries preferred gold to dollars. To stem an outflow of gold, Nixon also announced that, along with the controls and a program to stimulate the economy, the United States would no longer sell gold. This weak international position became a factor in the policy response to the energy shock, with U.S. appeals to oil-exporting nations to invest their surpluses here adding to the latter's sense that they were in the driver's seat.

As it turned out, 1972 was a great year economically. Controls kept inflationary pressure from producing higher prices, interest rates remained low, jobs were plentiful, and foreigners could not deplete U.S. gold stocks. The following year, the eve of the first oil price shock, was a different story. The inescapable rigidities of the controls produced mounting problems for the controllers, and the administration was forced to relax them at a time when the entire industrial world was caught up in an economic boom. Supplies of many commodities were tight as producers sought to build inventories of raw materials to assure their availability and to beat expected price increases. Meanwhile, other events produced a surge in food prices. It all seemed as if the Club of Rome's predictions of unsustainability of growth were coming true, much ahead of schedule.

Thus, when the Arab oil embargo was announced in late 1973, the world was already confronting tight markets and rapidly rising prices. That was part of the bad luck. In a calmer context the policy response to the energy price shock might well have been different and more effective. Perhaps there would have been less of a sense that the world was running out of everything and the marketplace could have been left to sort out winners and losers. Similarly, there might have been less worry that U.S. national security was threatened by a lack of oil.

Some policymakers welcomed higher oil prices. Former assistant secretary of state for business and economic affairs Thomas Enders once told me that Secretary of State Henry Kissinger actually welcomed the 1974 rise in oil prices. Kissinger, who clearly knew more about geopolitics than economics, based his Middle East policies on a number of assumptions, one of which, Enders said, was that the Saudi monarchy could not survive for more than five years. While not predicting that the monarchy would fall within that time frame, Kissinger meant to caution that it was unwise to assume that it would not. Thus, Kissinger decided that the key to stability in the region was likely to be Iran, and the rise in oil prices gave the shah the wherewithal to buy the military hardware his nation needed to play that role, Enders explained. While it cannot be confirmed, such thinking may be one reason that Kissinger, as Bohi and Darmstadter note, suggested that the United States was willing to consider price-setting agreements with commodity producers, reversing previous American policy.

We now know that there was never any actual shortage of oil either during the embargo period or in 1979, when motorists waited in long lines at gasoline stations. The fear of being caught short drove most oil users to hoard their stocks—including motorists, who substantially increased the average level of gasoline in their cars' tanks.

Bohi and Darmstadter say that the importance of this inventory response was "not appreciated when it was happening, nor is it understood to this day why stocks were built as they were, and prices responded precisely as they did" (Bohi and Darmstadter, 1994: 37). Perhaps that was true in 1974, but by 1979 some oil experts argued that uncertainty was producing hoarding and that spot shortages were due to the distribution system's inability to accommodate the sudden rise in apparent demand caused by the hoarding. Nevertheless, Energy Secretary James Schlesinger cited the leap in the disappearance of crude and products from primary storage in the first quarter of 1979 as evidence of profligate U.S. energy consumption. He did so to put pressure on Congress to pass energy legislation introduced by the Carter administration. In an ironic twist, members of Congress later accurately used data to show that there was in fact no shortage, and that oil company inventories rose in the first half of 1979. They argued that since there was no shortage, the price increases that occurred must have been the result of market manipulation by the oil companies.

Antipathy for the oil industry, shared by a large segment of the public, made it politically difficult for policymakers to rely on market forces to deal with the problems posed by the big jump in oil prices. The companies did not "deserve" higher prices for oil they had discovered before 1973; thus, a windfall profits tax was placed on the sale of such "old" crude. Since the major oil companies controlled much of the "old" crude, smaller independent oil companies argued that fairness required the big companies to share their windfall. Thus was born the entitlement program.

The fear of being seen as rewarding the oil giants while the price of gasoline and home heating oil was going up also contributed to a remarkable turnaround by the Ford administration in 1975 on the issue of oil price controls. President Gerald Ford began that year by imposing a $1 tax on imported crude and a comparable tax on refined products as part of an effort to force Congress to agree to end controls. Congress resisted, but eventually legislative circumstances developed that left decontrol within Ford's grasp. Unfortunately, his advisers convinced him that decontrol— estimates were that gasoline prices would rise by about 7 or 8 cents a gallon—would severely damage his chances of being elected in 1976. So Ford reversed himself and agreed to an extension.

The United States' weak international payments position also affected the response to the energy price shock. Well before the oil embargo, rising inflation in the United States undermined the international monetary system established under the Bretton Woods agreement of 1944 by calling into question its foundation: a fixed price of gold of $35 an ounce. In reality, this meant that the foundation was the dollar, and inflation shook the foundation. The energy price shock added fuel to the inflation fire and instability to the emerging system of floating exchange rates. At the same time, higher oil prices directly increased the U.S. trade and current account deficits and raised doubts that foreign capital could be found to finance the nation's external deficit without a massive increase in U.S. interest rates.

In the summer of 1974 I traveled with Shultz's successor at the Treasury, William Simon, on a Middle East trip whose primary stop was Saudi Arabia. Basically, Simon, hat in hand, asked the Saudis to help finance the U.S. external deficit by buying Treasury securities and depositing their excess cash in U.S. financial institutions. Meanwhile, questions were raised in other quarters as to whether the world's financial institutions were up to the job of recycling the oil exporters' surpluses. If they were not, some economists warned, higher oil prices, acting like a tax increase, could cause a worldwide recession or worse. A few prescient bankers noted that the intermediaries in the recycling, the banks, could be left holding the bag in the case of defaults by those to whom the exporters' surpluses would be lent.

This occurred in the summer of 1982, when the Mexican government became the first in a string of developing nations to announce that it could not meet its debt service obligations. This so-called LDC debt crisis left many of the largest U.S. banks actually, but not officially, bankrupt. To protect the U.S. financial system, which might have been threatened by the failure of any of these large institutions, Federal Reserve chairman Paul Volcker convinced the Reagan administration and other federal bank regulators that the banks had to be given time to earn their way out of bankruptcy. Thus, the regulators did not immediately require the banks to reflect the true value of the LDC debt on their books, which allowed the institutions to appear solvent even though they were not. The strategy worked, although later in the decade, when other losses mounted such as those from commercial real estate loans, some of the institutions were still in a weakened condition from their LDC losses.

While less direct than in the case of the LDC defaults, the energy price shocks were a major factor in the demise of the nation's savings and loan industry, which imposed huge costs on federal deposit insurance funds that had to be covered by taxpayers. The thrift industry, whose sole purpose was to finance housing, was predicated on maintaining low inflation. It borrowed short and lent long and was protected against a surge in its cost of funds by regulation of the interest rates it could pay to depositors. In a world of high inflation—due in part to the energy shocks—the thrift industry could not survive. This is not the place to chronicle all the mistakes the industry, its regulators, Congress, and the Reagan administration made that caused such large taxpayer costs. However, it is fair to say that had the second energy price shock not occurred, those costs might have been far smaller. Again, it was primarily a matter of poor policy. The price shock created most of the problem.

The energy price shocks also directly contributed to large losses by financial institutions in another way: high oil prices fueled a boom in Texas, Oklahoma, and other parts of the Southwest. When oil prices collapsed, so did the boom and real estate prices. Many financial institutions, including every large bank in Texas, went under in the late 1980s in the wake of losses on loans to energy companies and real estate developers. The eventual cost of the commercial bank failures proved to be manageable for the federal bank insurance fund. The thrift industry's losses in the Southwest were much higher, with most of the cost borne by taxpayers.

It is not much of a stretch to suggest that U.S. political history would have been considerably different in the 1970s and 1980s without the energy price shocks. For instance, if the second shock had not occurred, it is likely that President Jimmy Carter would have been reelected. He actually did not run that far behind Ronald Reagan in the popular vote. Absent the gasoline lines, high inflation, high interest rates, and the 1980 recession, Carter certainly would have fared better than he did. Had he had a second term, his approach to federal fiscal policy would have been very different from that of President Reagan. Carter would have pursued a much less radical approach to tax cuts and probably would not have had the political freedom to allow budget deficits to reach the heights they did under Reagan, even if he had wanted to. Tax cuts did occur, and deficits as a share of GDP did reach record levels. Spending choices, of course, played a role in the deficits, not just the tax cuts. The Federal Reserve responded to the loose fiscal policy with a tighter-than-normal monetary policy. This combination caused the value of the dollar and the country's trade deficit to soar.

It cannot be definitively claimed that the energy price shocks caused all of that to happen. However, an assessment of the economic impact of those shocks must consider the possibility that they made some otherwise manageable policy problems so bad that the U.S. political process could not manage them, causing nationwide suffering.

As Bohi and Darmstadter note, policymakers eventually concluded that the oil market had to be deregulated even if the result was higher inflation for a time. President Reagan, in his first month in office, made a great point of terminating oil price controls even though legislation phasing out controls had been passed three years earlier and the process was virtually complete before he was sworn in. Many other steps to dismantle the web of energy regulations came about later in the decade. Just how much energy receded as a policy issue is underscored in a collection of essays in *American Economic Policy in the 1980s*, published by the National Bureau of Economic Research (Feldstein, 1994). It contains essays on monetary, tax, budget, exchange rate, trade, LDC debt, and antitrust policies, policy toward the aged, and economic, health, and safety, and financial regulation. The only mention of energy is a reference to a late-1982 suggestion by Secretary of State George Shultz that a tax on imported and domestic oil could raise revenue to reduce budget deficits. Reagan proposed such a tax on a contingency basis: it would take effect only if deficits exceeded a specified level. The contingency tax was never enacted.

The story is not all that different in the 1990s, which may be the best evidence that Bohi and Darmstadter are right when they conclude that we really have learned a lot since the first energy price shock, though in many cases as the result of costly policy failures.

When Iraq invaded Kuwait in August 1990, the availability of oil supplies was jeopardized once more. Calculations (Verleger, 1993: 25) show that the response of world oil prices was similar to that in the price shocks of 1973 and 1979 for roughly the next thirty days. The path of prices in 1990 relative to their starting point actually was close to that of 1979 for about three months; then prices declined irregularly until, in January 1991, they returned to their starting point. U.S. military opposition to Iraq undoubtedly played a role in bringing prices down again because it limited the degree of uncertainty in the situation by making it clear that Iraq was not going to be allowed to invade Saudi Arabia or be left in a position to dictate policy to the Saudis.

Nevertheless, it is noteworthy that when the oil production capacity of two major exporters, Iraq and Kuwait, was removed from the world market, increases in output by the Saudis and others easily made up the difference, and prices returned to their former level. Other than military action, there was no rush by the government to interfere with the oil market. In fact, the Bush administration may have missed an opportunity to calm the initial market response by delaying for some months a sale of oil from the U.S. Strategic Petroleum Reserve (SPR). Nor were there sweeping claims of oil company profiteering, and most of all, no one seemed to be very concerned that the world would run short of oil.

One energy issue that was addressed in 1994 was whether to authorize adding more oil to the SPR, which stores nearly 600 million barrels in salt caverns in Louisiana and Texas. The Clinton administration wanted funds only for maintenance and improvements to the storage facilities, not to purchase more oil. On several previous occasions, Congress had insisted on continuing the SPR fill over the objections of the Reagan and Bush administrations. On September 30, 1994, President Clinton signed an appropriations bill into law (P.L. 103-332) that provided funds only for construction, operation, and maintenance of the SPR. Fill was suspended in fiscal year 1995 because of budget constraints, and to devote resources to refurbishment of SPR equipment and life extension of the SPR complex through the first quarter of the next century.

Verleger, like Bohi and Darmstadter, notes that the world oil market is far more open than it was twenty years ago, with many more producers, active spot and futures markets, and far less vertical integration of the oil industry (Verleger, 1993). Given those changes, and what has happened to world oil inventories over the past two decades, Verleger questions whether having the SPR actually makes much difference either to the oil market or to national security. As publicly held or mandated oil stocks have increased, private inventories have fallen, so that today the total is not much different from what it was in 1979.

Another significant difference from the response of the 1970s was the monetary policy adopted by the Federal Reserve in the fall of 1990. Federal Reserve policymakers believe that the central bank made major policy errors in the 1970s by not moving more vigorously to control inflation, particularly in the latter part of the decade, which was marked by rising inflation prior to the second oil price shock. In 1990, inflation was neither particularly high nor rising. The Fed had earlier raised short-term interest rates substantially to bring down an inflation that appeared to be accelerating in the context of a fully employed economy. More than a year before the Iraqi invasion, the central bank reversed course and began to bring rates down again as it sought what chairman Alan Greenspan described as a "soft landing," an approach that would curb inflation without putting the economy into recession. He now thinks that except for the invasion and its aftermath, the soft landing would have been achieved.

At a policymaking session in October 1990, Fed officials were divided over the best course for monetary policy, partly because the economic impact of the oil price surge was unclear. Greenspan and a majority of the policymaking group, the Federal Open Market Committee, did not want to take any chance that the impulse from rising oil prices would become embedded in the nation's wage and price structure. If this meant a decline in real output, then that was preferable to letting inflation rise in the short run only to have to sacrifice output and employment to deal with inflation later.

Some economists are still puzzled why the 1990–91 recession occurred. One suggestion (Hall, 1993) is that there was a spontaneous decline in consumer spending. If that was the case, the only plausible reason for such a behavior change was rising oil prices and the memories they recalled of the difficulties of the 1970s.

The Fed's refusal to lower interest rates more swiftly in the fall of 1990 and during 1991 was sharply criticized by President George Bush and several senior administration officials, particularly Treasury Secretary Nicholas Brady, and many members of Congress. Greenspan has argued that had the central bank brought short-term rates down more rapidly, financial markets—mindful of the short burst of extra inflation set off by rising oil prices—would have reacted by raising long-term interest rates as inflationary expectations mounted.

Conclusions

I agree with Bohi and Darmstadter that the events of the 1970s could not be repeated today because of changes in markets and past policy mistakes that are not likely to be repeated. I also agree that our ability to forecast energy prices and consumption remains woefully inadequate.

Recent actions by the International Energy Agency amplify this point (IEA, 1994). Compared with its 1993 assessment, the IEA has raised its prediction of world energy and oil consumption for 2010. Nevertheless, it has lowered its oil price projection by $2/barrel. In its 1991 report, the IEA projected that oil prices (in real 1993 terms) would rise to $38/barrel by 2000. Last year that figure was lowered to $30 in 2005. The IEA now predicts $28. One reason for the lower figure is "greater optimism on the part of industry regarding improvements in technology. For example, it is now reported that a large part of Canada's resources of tar sands and *in situ* bitumen could be economical at or below such a price level. Many other currently expensive forms of energy could also become economical at an oil price range between $25 and $30" (OECD/IEA, 1994: 25).

While driving on I-70 in western Colorado on my way to Salt Lake City recently, I saw a scattering of homes that were to be the forerunners of thousands more along Battlement Mesa. They were to house employees of a burgeoning oil shale industry a few miles further west in Parachute Creek, where Exxon was constructing a massive project. In 1981 Exxon chairman Clifton Garvin came to lunch at the *Washington Post* and discussed his plans for the oil shale project, which he expected to produce millions of barrels of oil daily. He said the nation would need at least 9 million barrels a day from the entire oil shale region of northwestern Colorado. This oil was so desperately needed for the nation's economic health and national security, he argued, that that corner of Colorado should be declared a national energy zone in which normal environmental restrictions did not apply. Other company executives with him were taken aback at that suggestion, and Exxon never publicly pursued the idea. Only a year later, Exxon abandoned the project because of high costs and falling oil prices.

Maybe tar sands will be viable at $25/barrel, or maybe not. Everyone's record on forecasts is pretty terrible. As the IEA candidly admitted:

Oil price has been extremely difficult to predict. The IEA (and to a large extent these figures represent the industry consensus) thought it would stay flat or go down when in fact it went up, then they thought it would stay up or rise further when in fact it came down. And these errors were very large. Over the whole period [i.e., since 1974, when IEA was founded], price forecasts in real 1993 dollars range from $10 to $70 per barrel. . . .

Concerns in the 1990s have . . . moved on. There is less concern about shortages: the supply problems envisaged in the 1970s have not materialized, more oil and gas has been discovered and probably will continue to be discovered in the future. The ultimate resource constraint is currently regarded as too distant to be a major policy concern. There is greater confidence in the market. Although many countries have policies which are relevant to energy efficiency, these policies are not primarily motivated by global energy shortage. Security of supply, which was a major concern in the 1980s has moved down the policy agenda, and the environment has become a key concern. (OECD/IEA, 1994: 286–87)

Currently, the average price of gasoline at the pump in the United States, including all taxes, is the lowest in real terms that it has ever been. Who in 1974 would have thought this could possibly have been true in 1994? That is one measure of how far we have come in the past two decades. So where are we headed? The reader's guess is probably better than mine, but based on the record, we are both apt to be wrong.

REFERENCES

Bohi, Douglas R., and Joel Darmstadter. 1994. *The Energy Upheavals of the 1970s: Socioeconomic Watershed or Aberration?* Washington, D.C.: Resources for the Future.

Feldstein, Martin, ed. 1994. *American Economic Policy in the 1980s.* Chicago: University of Chicago Press.

Hall, Robert. 1993. "Macro Theory and the Recession of 1990–91." *American Economic Review* 83 (2): 275–79.

Meadows, Dennis, et al. 1972. *The Limits to Growth.* New York: Universe Books.

Organization for Economic Cooperation and Development, International Energy Agency (OECD/IEA). 1994. *World Energy Outlook.* Paris: OECD.

Verleger, Philip K., Jr. 1993. *Adjusting to Volatile Energy Prices.* Washington, D.C.: Institute for International Economics.

PART TWO

UNRESOLVED ISSUES AND THE ROLE OF POLICY ANALYSIS

What role does the policy analyst play in shaping energy policy? Have policy analysts learned to do a better job of explaining energy-related behavior and predicting future energy supply and demand? And can policy analysis help decision makers more effectively distribute the costs, benefits, and risks of energy throughout society? Finally, just who are these policy analysts, and what do they believe?

The four chapters in this part are held together by an important thread: the notion that the basis for sound policy is sound information and analysis. Without it, decision makers grappling with critical energy decisions have little hope of making the right decisions. With it, policymakers may still fail if the information and analysis are unclear, incoherent, inconsistent, or poorly communicated.

Harvard University's William Hogan argues that despite the inherent problems of large energy modeling or forecasting efforts and their associated analyses, these attempts to explain the behavior of energy producers and consumers have taught us much about energy policy since 1973. Their basic contribution has been to further our understanding of the motivations for, and constraints upon, energy decision making. Hogan contends that models may offer good insights that can be successfully communicated to policymakers; they may offer good insights without good communication (a "first order" failure of the policy analyst); they may fail to predict; or they may be incorrectly applied.

The rejection of Project Independence in 1974; the successful estimation of supply and demand elasticities in energy markets and their implications for the relationship between energy use and economic growth; analyses of the effects of price and allocation controls on energy markets; and analyses of competitive forces on natural gas and electricity transportation markets despite the apparent presence of natural monopolies—these are all the result of good analyses well communicated.

Economic models rightly predicted that removing price controls from natural gas would hasten exploration, production, and distribution and would ensure a more stable, abundant supply. Hogan feels that the establishment of a Strategic Petroleum Reserve (SPR) was a bad decision based on incorrect forecasting. He contends that the SPR has been little more than an unnecessary drain on the federal budget.

The American Petroleum Institute's Michael Canes questions the degree to which energy predictions have led to better policy. Contending that *effectiveness* is not easy to define, Canes suggests that despite the abandonment of Project Independence, other national and state policies may be designed to produce the same outcome. Such policies include subsidies for alternative fuels (e.g., in California and Texas) and requirements that utilities buy power from independent power producers—despite problems in the availability and reliability of such power.

Most important, says Canes, previous predictions may have been wrong for reasons lying entirely outside the framework Hogan introduces: exogenous political variables (e.g., pressures upon Congress) and changes in technology may also affect policy. Moreover, forecasted changes in real energy prices alone are not very good predictors of technological change. As Canes concludes, failure to follow advice may not be due as much to misunderstanding as to bona fide disagreements over desired outcomes.

Glenn Schleede, of Energy Market and Policy Analysis, Inc., concurs with this analysis but adds a new wrinkle. To Schleede, few decisions by government in the area of energy policy are ever motivated solely by rigorous modeling of outcomes. Too often, myriad programs designed to promote federal government support for energy R&D, to subsidize one energy source or another, or to tax some form of behavior are motivated at the behest of private interests. Such subsidies merely exacerbate public distrust of government's ability to fix energy problems.

Oak Ridge National Laboratory's David Greene examines the continuing impacts on American society of OPEC and of the 1973–74 embargo. The embargo and its aftermath have forced the United States to transfer considerable portions of its national wealth overseas, taking away capital that could have been invested in improving the quality of education, health care, the environment, jobs, and our children's future. Suggesting that energy markets alone cannot resolve the problem of oil dependence or encourage more efficient energy use, Greene concludes that some interventionist approaches, such as automobile fuel-economy regulations, may not only be necessary, but their previous use may have averted worse price shocks. Mandated vehicle fuel-efficiency standards, renewed competition in the electricity generation industry, and oil and gas price decontrols—all of which would probably not have come about without the energy crisis— have helped to ensure a more stable energy supply and greater institutionalization of energy efficiency.

7. Predictions, Prescriptions, and Policy

Lessons from the Record

WILLIAM W. HOGAN

Large modeling or forecasting efforts and their associated analyses are inherently problematic. I argue that despite these problems, energy modeling and analysis efforts conducted since the energy shock of 1973 teach us much about energy policy. The most common and largely accurate criticism of much of formal modeling was summarized by Amory Lovins, a prominent energy analyst who followed a different path:

Such models have trouble adapting to a world in which, for example, real electricity prices are rapidly rising rather than slowly falling as they used to. . . . Extrapolations have fixed structure and no limits, whereas real societies and their objectives evolve structurally over decades and react to limits. Extrapolations have constants, but reality only has slow variables. . . . Extrapolations assume essentially a surprise-free future even when written by and for people who spend their working lives coping with surprises such as those of late 1973. Formal energy models can functionally be stripped of surprises, but they can say nothing useful about a world in which discontinuities and singularities matter more than the fragments of secular trend in between. Worst, extrapolations are remote from real policy questions. (Lovins, 1979: 64)

Lovins's alternative to formal modeling is an analytical approach fraught with difficulties. It rejects exploitation of available data in favor of more speculative ruminations in which it is difficult to separate prescriptions from predictions that can be understood and reproduced by others. Therefore, despite its limitations, formal modeling is often better than the alternatives.

It is easy to show the follies of forecasting, and energy forecasting has a well-known record of mistakes and failures. As the famous physicist Niels Bohr is reported to have said, "It is difficult to make predictions, especially about the future." For example, energy forecasters are painfully familiar with previous published energy demand projections in which total energy demand declined at a compound rate of 7 percent per year between 1972 and 1983 (Greenberger and Hogan, 1987: 256). This phenomenon was repeated in countries other than the United States. In the electricity sector, the infamous National Electric Reliability Council (NERC) "fan" captures the failed extrapolations of electricity demand starting in 1973 by displaying each subsequent year's NERC projection shifted one year later to give the appearance of an unfolding fan, summarizing the stubborn belief that old electricity growth trends would repeat (Nelson, Peck, and Uhler, 1989).

A more interesting set of issues that addresses the connections between predictions, prescriptions, and policies are the complicated interactions between forecasting, analysis, and the political battles among interest groups. Previous analyses of the 1970s, and the rush of large-scale analytical efforts (Greenberger, 1983), show that there is no tidy, linear progression from formal model to final policy: "Professors of policy analysis would like to tell their students that an analysis done professionally attracts attention and gets used. We found, instead, that the highest quality analyses received the least attention and exerted the lowest immediate influence on policy. What is important is 'timing . . . ,' clarity, and whether the results accord or clash with the beliefs of those in power" (Greenberger and Hogan, 1987: 260).

A clash of values and beliefs can obscure the impact and importance of analysis. Few energy analyses that have assaulted existing values and beliefs have had wide influence. Amory Lovins's work may be the exception that proves the rule (Greenberger and Hogan, 1987: 260). But even he worked tirelessly to repeat and hone his message.

Wide acclaim and recognized success for forecasts are unlikely. If the answer is counterintuitive, it is wrong. And if the answer is counterintuitive, the most likely cause is a mistake in the analysis. If so, everything that can be done should be done to uncover the mistake. If the mistake is found and corrected, the conventional, intuitive answer will be shown to be right. If an explanation is found that supports the surprising answer, intuition will change and the decision rule still applies. This latter case is where analysis makes its greatest contribution. Whatever the outcome, the intuition, and not the analysis, is likely to get the credit. A good answer had better be consistent with common sense.

An interesting domain for studying the role of formal modeling is the interaction of insight from the analysis and the clarity of communication in the hard cases where "results . . . clash with the beliefs of those in power." In general, the message of the last two decades is simple: analysis must provide insights more than predictions, and the insights must be communicated (Geoffrion, 1976; Huntington, Sweeney, and Weyant, 1982: 449–62). Marketing is just as important as production.

Applying this simple rule, however, is far from trivial. Its application has important implications that go far beyond a preference for simple models over complex ones. Analytical attention that focuses on insights and their communication can determine both the form and the content of the analysis. The following case studies are illustrative of past successes and failures in, and future opportunities for, acquiring and communicating the insights of formal analysis and energy models.

Insights Acquired and Communicated

The first domain to illustrate success is the set of examples from modeling and analysis. Here *success* is defined as cases where the conventional wisdom changed from one view to another, and in the same direction as indicated by available predictions and prescriptions. This definition forgoes a claim of success when the analysis merely confirms the conventional view. But this definition of *success* is generous in that it begs the question of what else could explain the change in the prevailing conception. The case studies summarized here do not prove, but they do support, the assertion that the analysis contributed to redirecting policy. These successes share a common theme of repeating the argument until the new idea permeates policy. There are few if any examples of analysis and modeling producing a sharp turn in policy. From this perspective, the educational function of policy analysis and modeling looms large.

Energy Independence

The shock of the oil interruption and price increases accompanying the 1973 Yom Kippur War precipitated a major effort to review and redirect the nation's energy policy. President Richard Nixon, looking for dramatic impact—and a diversion from Watergate—set the goal of energy independence. Project Independence was born on November 25, 1973, in a televised speech on energy by President Nixon. The president carefully avoided setting a goal of zero energy imports: "What I have called Project Independence—1980 is a series of plans and goals set to ensure that by the end of this decade Americans will not have to rely on any source of energy beyond our own. . . . The capacity for energy self sufficiency is a great goal" (Havemann and Phillips, 1973: 1637).

Despite its cautious language, the goal of energy independence was widely interpreted as calling for zero oil imports. At least one early analysis in the Department of the Treasury led to a plan that started from the assumption of zero imports in 1980 and worked backwards to a massive program of government subsidies to expand domestic production. However, the Project Independence analytical effort soon challenged the conventional view and proved less ambitious. Eric Zausner, who supervised and polished Project Independence's analysis, recalled that his task amounted to two simple redefinitions of President Nixon's original goals: "redefining 'independence' and redefining '1980'" (Havemann and Phillips, 1973: 1637).

The message that emerged from the supporting studies and computer models was that the goal of literal energy independence was ill conceived and would be prohibitively costly. At best, an aggressive program put in place in 1975 might eliminate oil imports for a short span around 1985, but the respite would be brief as natural petroleum supply depletion and energy demand growth drove the United States inexorably toward more oil imports. A more plausible and useful goal, embodied in the final Project Independence report, would be to reduce oil imports through a combination of programs to increase energy efficiency, enhance domestic production, and better prepare for what were expected to be new oil supply interruptions.

The analytical work done in support of the report continued to be applied and expanded, but the basic policy theme was permanently changed. Huge programs like the $100 billion Energy Independence Authority, to subsidize massive capital investment in the domestic energy industry—originally developed by Vice President Nelson Rockefeller and later proposed by President Gerald Ford (*Public Papers of the Presidents,* 1975: 1494)—fell away before analysis showing that such subsidies could not achieve independence and would be less cost-effective than other modest alternatives to increase efficient use of resources.

The policy analysis work spawned by Project Independence was never fully embraced, and there were many skeptics: "*DOWNGRADED:* FEA will continue to issue its annual updates to the Project Independence report, but FEA Administrator John O'Leary says he doesn't place much importance in them. Econometric studies applied to mineral fuels 'are not worth the paper they are written on,' O'Leary claims" (*Energy Information,* 1977: 4).

Despite flaws in the models and the analyses (Hollaway, 1980), their main ideas could be communicated, tested, and understood. Moreover, it was not the "econometric studies" or associated models that were the principal product, but the insights acquired from them. Domestic oil supplies were becoming more and more expensive; oil imports were relatively cheap; coal was available but hard to use in a cost-effective and environmentally acceptable way; and the market would adjust to eliminate any gap between supply and demand. These ideas changed the definition of *energy independence* and led to the quiet demise of the original target date of 1980.

When compared with the conventional wisdom embodied in Nixon's speech, the fact that these ideas now seem obvious is testament to the power of the policy analysis completed, repeated, and communicated under Project Independence's analysis.

Supply and Demand Elasticities and Economic Growth

Introduction of prices and price elasticity into the policy debate was a premier analytical contribution. It captured the importance of markets and the ability of the economy to adjust to the changing availabilities of energy. Before Project Independence and related studies, prices barely entered the analyses. The dominant form of modeling was the extreme version of extrapolation condemned by Amory Lovins. The desired growth of the economy would lead to an inexorable growth in energy demand. Left alone, sources of domestic supply would decline, especially scarce oil and gas resources. The neo-Malthusian difference between supply and demand was a gap that called for government intervention to subsidize and expand secure energy supplies.

To the economist, the flaw in the logic was self-evident. The gap between supply and demand would never appear. Gaps could exist only briefly. When supplies fell short of demand, prices would rise in the natural rationing mechanism of the market. Higher prices would curtail demand and increase supply, soon eliminating the gap. Hence, from the economist's perspective, the question was not how to mobilize the government to provide energy independence. The proper focus was to look for and eliminate market failures that might prevent this natural process or increase the cost of adjustment.

The debate between those who saw impending gaps and those who saw opportunities for markets to adjust has been well documented (Greenberger, 1983). It spawned several investigations emphasizing the demand side with a variety of energy models to narrow the range of disagreement about the size of the aggregate elasticity of energy demand (Energy Modeling Forum, 1977; 1981). There was an equally vigorous debate on the comparative advantages of econometric studies of historical data versus engineering analyses of individual energy conservation technologies.

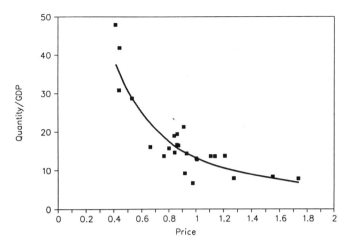

Figure 7.1   OECD Gasoline Demand, 1973
Source: Economic Report of the President, 1975.

Resolution of the argument over prices and market adjustments is now fully embodied in the conventional wisdom (Hogan and Manne, 1977; Hogan, 1979). In the short run, price elasticities are small because there is little room to adjust energy use, and sharp shocks to supply and demand can cause large movements in price; hence the problem of oil supply interruptions. But with the slow turnover of capital stock, there are large opportunities to exploit new technology and change the utilization of capital, labor, energy, and materials. Over a reasonable period of time, adjustments can be large enough to make energy scarcity a lesser problem for all but those employed in the most energy-intensive industries.

The necessary evidence was readily available at the time of the great price-elasticity debate. One of the most striking examples of the long-run ability to adjust to different prices is also one of the simplest. Data shown in figure 7.1 compare the intensity of gasoline use per unit of GDP across OECD countries in 1973, plotted as a function of the final price of gasoline. A variant of this figure first appeared in the *Economic Report of the President* in 1975. The argument was that the complex, confounding analytical effects of price shocks can be avoided by looking at cross-national comparison after a period of relative stability of oil and gasoline prices. From this perspective, the figure depicts the long-run equilibrium of gasoline demand, after all the adjustments have been made to accommodate different gasoline prices. The implied elasticity of gasoline demand is near unity, consistent with many later studies that were completed with more data.

Figure 7.1 is as close to the textbook example of a highly elastic demand curve as we are ever likely to find.

Contrary to the initial fears and assumptions of the early calls for massive energy subsidies to maintain economic growth, the consensus of the analyses was that long-run demand elasticities were really quite large. Hence, economic growth could proceed almost indefinitely without corresponding growth in energy use. If energy became scarce, technology driven by higher prices would increase real economic output without using more energy. Thus, massive subsidies to the energy industry were viewed as detracting from economic growth, rather than necessary to its maintenance. Steady repetition of the many energy models and analyses was an important part of this story.

Price Controls and Shortages

A similar consensus emerged from analyses of supply-side responses to higher energy prices. There may not have been much confidence in the details of individual econometric models of oil and gas supply, but the evidence of positive price response was compelling. However, in each case, evidence had to overcome conventional wisdom through repeated studies and hard experience.

Consider the case of natural gas. By the mid-1970s extensive price controls had produced increasing shortages of natural gas in the interstate market. The excess of demand over supply caused growing curtailments that closed schools and factories in order to allocate the scarce supplies of available natural gas to higher-priority residential customers. By 1978 the Powerplant and Industrial Fuel Use Act forbade the use of natural gas in new power plants in an attempt to conserve limited supplies of natural gas for higher-valued uses. At the same time, the Natural Gas Policy Act imposed a complicated set of price regulations and ceilings designed to extract the few remaining gas reserves from deep deposits without allowing prices to increase much for the supposedly fixed supply of traditional sources of natural gas.

The result was a predictable fiasco. Price regulations distorted the industry completely out of connection with the underlying economics and created the anomalies of high-priced surpluses and subsidized consumption. This in turn drove up total demand. An accounting of the flaws in the argument and the high cost of policy recovery is now available (Pierce, 1988: 1–57; 1993). The failure was to approach natural gas as something intrinsically different from other commodities, with large segments of the market

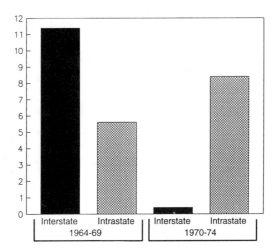

Figure 7.2   **Average Annual Reserve Additions of Natural Gas, 1964–1969 and
1970–1974 (trillions of cubic feet)**
Source: U.S. FEA, 1976: 122.

assumed to be impervious to the incentives price. As later studies showed,
the opposite was closer to the truth. Natural gas supply and demand *was*
responsive to price incentives. With unregulated prices the market would
balance itself, often at much lower prices than those imposed by rigid
regulations.

The evidence was readily available before passage of the Natural Gas
Policy Act in 1978. Figure 7.2 is from the *National Energy Outlook* of 1976,
prepared by the U.S. Federal Energy Administration. Before 1970, natural
gas was readily available. By 1970, curtailments began to appear, but only in
the interstate market. The explanation was simple. By the early 1970s, prices
for interstate sales were capped by controls at "52¢ per thousand cubic
feet. . . . In contrast, unregulated prices for gas sold on the intrastate market
usually range between $1.00 and $1.50 per Mcf" (U.S. FEA, 1976: 117–19).

As shown in figure 7.2, the market responded with ample supplies for the
intrastate market and a complete collapse of new supplies in the regulated
interstate market. It would be hard to find more dramatic evidence of the
supply-side effect of prices. However, it took more than a decade and great
cost to convert the analytical argument into natural gas policy. By the early
1990s the old analytical heresy that natural gas is a commodity for which
supply and demand respond to price became the new conventional wisdom. Price controls create shortages, and market competition restores a
supply and demand balance.

A similar and slightly faster revolution took place in oil markets analysis. However, the notion of a supply response to higher prices was still difficult to translate from analysis to policy. Coupled with the infamous "entitlements" program, domestic price controls in 1973 subsidized oil imports and increased oil dependence (Kalt, 1982). The only shortages that arose did so because of federal rules that allocated supplies to local markets but did not keep up with shifts in local demand. It soon became clear that the gasoline lines were caused by government intervention, and price controls only exacerbated the problem. Eventually, President Jimmy Carter took the political heat and used his authority under the Emergency Petroleum Allocation Act of 1973 to gradually eliminate domestic price controls on oil. In his first official act, President Ronald Reagan used the same authority to complete the process of price decontrol a few months early. Thereafter the Reagan administration felt able to take full credit for the coincidence that world oil prices declined almost immediately.

There has not been a full tabulation of the costs of oil and gas price controls. However, it is clear that the many supporting policy studies, coupled with the experience of shortages and unwanted costs, convinced most analysts and policymakers that the country would be better served by exploiting the market and allowing the evidently large long-run supply and demand elasticities to do their work. The precision of the estimates was never as important as the coherence of the argument and the reinforcement of the models in explaining the evolving turmoil. No single study turned the policy direction. The effect was cumulative. The analytical framework of price-responsive supply and demand came to replace the neo-Malthusian view of the irrelevance of price incentives. Energy pricing and market policy moved from protecting favored groups, or preventing wealth transfers, to promoting competition and free markets.

Natural Monopolies and Competition

Perhaps the most striking example of the power of analysis and the ability of an economic argument to affect the design of policy is the treatment of natural monopolies' prominent role in the energy market. Once convinced of the power of the forces of supply and demand, the competitive market is usually recommended as best able to find the efficient balance. A fair amount of evidence supports the notion that supply and demand respond to price changes and that the market can achieve an acceptable equilibrium.

In the energy market, however, some segments come close to the definition of natural monopolies. In the case of natural gas, for instance, the interstate pipeline bottleneck separated the many customers from the many producers. For competition to operate, so the argument went, customers and producers had to reach each other through the pipeline. Pipelines are a natural monopoly and an essential facility. Government intervention would be required to guarantee "comparable access" for all participants (Kalt and Schuller, 1987).

The opposite view was that the pipelines had effective competition or would be compelled under regulation to make cost-effective purchases of natural gas. Thus, no further government intervention was required. By its very nature, there is likely to be scant evidence of how the market would operate if there was no regulated monopoly. In most cases of access to an essential facility, the argument is based on analogy or first principles rather than empirical evidence.

Analysis from first principles and argument by analogy were applied to natural gas. Appeals to the experience with airlines, trucks, and telephones were made to bolster support for the benefits of competition and the need for access to essential facilities (Maryland People's Counsel, 1984). The monopoly pipeline that transported gas to others and sold gas on its own could hardly be expected to avoid the temptation to control one for the profit of the other. Examples of real or imagined discrimination accumulated as the debate progressed, but the touchstone was always the analytical idealized level playing field.

The notion of nondiscrimination in pipeline transport conveniently dovetailed with the simple model of a pipeline viewed as a bundle of straws connecting producing fields to the final consumer. Although flawed in detail, the basic notion that every customer could have its own little bit of pipeline capacity, reaching all the way back to the producer, survived intact in the final rules. Transition cost was high (Pierce, 1988; 1993), but federal rules embodied in Order 636 of the Federal Energy Regulatory Commission (FERC) completely reoriented the natural gas industry away from regional segmentation and interstate dominance by a few pipelines to a true commodity market where the only regulated service is transportation. As a result, the functions of buying and selling gas have become a fully competitive, highly integrated market (Doane and Spulber, 1991).

The same process that unfolded for natural gas extended to the electricity market in the early 1990s. Successful access to natural gas pipelines was used as the analogy supporting the need for access to the integrated transmission network connecting potential sources of electricity to customers wanting to take advantage of competition (Pierce, 1988; 1993). The Energy Policy Act of

1992 (sec. 211) gave FERC authority to ensure nondiscriminatory access to the electric transmission network in support of development of a fully competitive wholesale market.

The difficulties FERC faces in electricity are greater than those in natural gas because of the inherent complexity of the electric system. The simple model of a bundle of straws does not apply here. Something else must take its place. Competing generators must have access to the essential facilities that stand between them and potential wholesale market customers. Transmission is the essential facility, and open access to it is necessary for development and operation of a competitive market.

While access is necessary, however, there is more to transmission than simple connection to wires. The free-flowing electric grid requires coordination of short-term operations to maintain system stability and achieve least-cost dispatch. This coordination function operates most efficiently through a power pool that provides many services implicit in the economic dispatch. Dispatch provides an automatic source of backup supplies, short-term excess sales, reactive power support, spinning reserve, and many other services bundled into electric power transmission. Without equal access to these functions, new market participants will be at a competitive disadvantage relative to those having access to the full array of power-pool benefits. However, there is no consensus on how to provide this comparable access.

Learning from the gas experience, in 1993 FERC launched a series of inquiries aimed at analysis and investigation of alternative models for industry organization and access. These hearings steeped FERC in the details of economic incentives and engineering practice. While the debate was not resolved by late 1994, analysis and economic models stood at the center of the argument. The modelers showed that they had partly learned the lesson of the last two decades. The models offered in these FERC proceedings emphasized concepts and insights in preference to numbers and forecasts (Hogan, 1992: 211–42; 1993a: 171–200; 1993b: 18–29; 1993c). If the analysis is to affect electricity policy, which is still open to question, success will depend on finding a simpler way to capture the insights and relate them to the real world.

Insights Acquired but Not Accepted

We turn from successes, generously defined, to failures by any definition. Here the analyses failed to overcome conventional wisdom, even though the conventional wisdom may have been wrong. Something was awry. At a minimum, the insights seen by the analysts were not successfully communicated or accepted.

Emergency Management and the Strategic Petroleum Reserve

In 1973 the sudden disruption of the oil market and the associated price rise caused the United States and other oil-importing countries to reduce import dependence and stabilize supplies. Proposals for new policies ranged from promotion of energy conservation to reduce consumption to creation of a large synthetic fuels industry to convert abundant coal into newly scarce oil. In virtually every case, the debate over policy merits and the need for government intervention was contentious and confused.

The principal exception to this divisiveness was virtually unanimous support for an emergency petroleum storage capacity. The early analysis in the Project Independence report estimated the cost of a sudden oil supply interruption as a short-term reduction in GNP of $33 billion for each million-barrel-per-day supply curtailment (U.S. FEA, 1974: 377–92). The benefit of avoiding this loss was weighed against the ten-year cost of buying and storing crude oil, chiefly in salt domes, to be available for quick use in an emergency. Use of the ten-year horizon implied that such an interruption would occur once a decade, a forecast consistent with market disruptions and oil price increases since 1979. Borrowing from the wisdom of Joseph, who advised the pharaoh to store grain in times of plenty to prepare for times of famine, the analysis supported a large emergency reserve in excess of 1 billion barrels.[1]

Later work, notably that of Teisberg (1981: 526–46), applied principles of dynamic programming and game theory to reinforce this basic analysis (Hogan, 1983). These and many other studies concluded that the United States and other oil-importing countries should create, maintain, and, when necessary, use large supplemental oil supplies (Weimer, 1982; Leiby and Russell, 1989). The resulting Strategic Petroleum Reserve (SPR) became the centerpiece of U.S. energy security policy. Although the SPR was authorized by Congress and included in the budget, the powerful Office of Management and Budget (OMB) dissented. OMB noted that the cost of the SPR came out of the current budget, and the benefits would only come later. Because of this, progress was slow and Congress pressed to have the reserve filled (U.S. GAO, 1983; Stanfield, 1985: 1024–26). Eventually, OMB objections were overcome and the SPR grew slowly, reaching 587 million barrels in July 1990, just before the outbreak of the Persian Gulf War (U.S. DOE/EIA, 1992: 45).

Saddam Hussein's invasion of Kuwait removed Iraqi and Kuwaiti oil from the international market and in a stroke eliminated the excess capacity that had plagued OPEC with an oil glut. The predictable result was a sharp jump in oil prices, from $16/barrel in July 1990 to $30/barrel in September 1990 (U.S. DOE/EIA, 1992: 105). The U.S. economy was threatened, and thus was presented with the first major opportunity to apply the SPR and gain some of the benefits that came with the expense of filling the reserve during a high-cost era. There were immediate calls for using the SPR to supplement imported oil.

The SPR sat safely in its domes of salt, unused until January 1991, at the initiation of Desert Storm. Oil prices fell almost immediately, more as a result of the successful missile attacks shown live on the Cable News Network from downtown Baghdad. Early military success guaranteed the security of Saudi Arabian oil and changed perceptions and power relations in the world oil market. The SPR arrived too little and too late.

What went wrong? One concern in the early analysis of the SPR was that the very act of using the reserve could detrimentally signal to the world oil market that a developing problem was a true crisis. Thus, the president would be reluctant to acknowledge the severity of an oil market interruption in hopes of avoiding panic and to allow time for diplomacy to work. This early concern produced a variety of proposals to insulate the president from the immediate decision, including financial options and regular sales from the SPR. As serious as this weakness in the SPR process may have been, however, it hardly explains the reluctance to use the oil in the early fall of 1990. By then troops were on the ground, and President George Bush was working vigorously, both publicly and privately, to assemble the successful allied force that launched a major war to recover Kuwait and isolate Iraq. Arguably, a decision to use the SPR could not be more serious than a massive commitment of armed forces. This setting therefore reinforced both the need and the opportunity to use the SPR without the possibility that its very use would be counterproductive.

In retrospect, SPR policy analyses failed the test of "clarity," and "whether the results accord or clash with the beliefs of those in power." The wisdom of Joseph and the pharaoh was misconstrued in the very name of the SPR. The president and other senior decision makers believed that, like Joseph's grain, the reserve stock should be saved until it was really needed—the last source of supply rather than the front line of defense. But this missed the whole point of the underlying analysis in every SPR study: that the benefits came from avoiding the early economic shock effects of the sudden price rise, not from a prolonged capability to sustain a high level of imports. The role of the oil reserve was to cushion the transition, not to be saved till last. More than one wag suggested that it would have been better if Congress had created a *Tactical* Petroleum Reserve to convey the idea that oil should be drawn down quickly.

This analytical fine point was not so fine as to be missed in the choice of storage for the reserve. The choice of salt domes over other alternatives, such as shut-in production capacity, followed from the need to provide a very rapid flow rate once the reserve was opened. Conventional oil wells produce too slowly, and the SPR "wells" had to be the most prolific in the world when in use. This need for rapid production derived naturally from the analysis that showed that the benefits of SPR use came from early, rapid withdrawal, not later leisurely use.

The need for rapid withdrawal was lost in implementation, both in the United States and among its allies' diplomats, who were virtually unanimously in favor of saving supplies for a real emergency, all the while conducting one of the largest military buildups in history. It is hard to imagine a more ideal circumstance for using the oil reserve, and we did not. Given this experience, it is difficult to justify continuing the reserve without at least communicating its value, changing the policy for use, and changing its name.

Energy Security and Oil Import Fees

The costs of oil import interruptions that led to nearly unanimous support for the SPR did not produce agreement on the need for further intervention in the form of an oil import tariff. As a protagonist in the debate, I view the issue with some bias. However, it is not necessary to take a position on the merits of the arguments to make the case that any insights that spring from analysis have been lost in the conflict of ideology and interests.

The case for an oil import tariff is simple in the abstract (Hogan, 1981). If higher oil imports lead to higher prices and greater exposure to the damage of oil supply interruptions, then associated costs are not priced in the marketplace. These externalities create a wedge between the social and private costs of oil imports that could be replaced by an oil import tariff.

The case against an import tariff is equally simple. Free trade is generally good. If the larger objective of free trade, such as the General Agreement on Tariffs and Trade (GATT), is jeopardized by a tariff targeted at oil, then the balance can be tipped against market intervention. Besides, the government can ruin any good idea; witness the SPR blunder.

This summary of the case against an oil import tariff is different from the usual arguments:

—Everyone knows that tariffs reduce economic welfare; it is in the text-books. (The textbooks soon used OPEC and oil as examples of exceptions to the rule.)

—An oil tariff is protectionist and creates gains for domestic producers. (True, but irrelevant in simple aggregate welfare analysis. Gains to holders of domestic oil reserves—e.g., Harvard, California pension funds—are not losses to the nation.)

—A tariff will reduce macroeconomic output and lower economic welfare (U.S. DOE, 1987). (Probably wrong, but if the analysis holds up, it argues symmetrically for subsidies for oil imports, but subsidies are also approved. There is no analytical support for this asymmetry.)

In short, opposition to an oil tariff has been built on ideological arguments isolated from the reality of the world oil market, which has been dominated by a cartel and for which the usual rules of thumb do not apply. Yet these arguments have been successful, and attempts to make the opposite case have failed. Failure can be partly attributed to inability to find the right model or metaphor to overcome the "clash with the beliefs of those in power." Free-market arguments applied to the competitive domestic market were transferred uncritically to the cartelized international oil market.[2]

Arguments against an oil tariff appealed to easy slogans and familiar truths. The arguments for an oil tariff built from unfamiliar perspectives and no appealing model. The notion of "internalizing the externalities" was no match for that of "defeating protectionism." The simultaneous call for removing domestic price controls and imposing an import tariff was viewed as inconsistent rather than as a coherent theory about the role of government based on a recognition of the degree of market failure. If there was any insight in the analysis, there was a failure of communication.

Insights Acquired but Not Yet Communicated

In addition to the successes and failures of communicating insights, energy market analyses have an inventory of important, evolving ideas that have not been rejected but have yet to be understood and communicated. They remain challenges to analysts and policymakers. Two examples illustrate the difficulties in finding good ways to analyze and communicate in the presence of hard problems.

Technological Change and Energy Demand

Over short time periods, technology is fixed and the economic focus is on choice and substitution. The analysis of energy demand can be reduced to the investigation of price incentives and opportunities to substitute capital, labor, and materials for energy at the margin. This framework dominates most models, explicitly or implicitly, in the form of fixed demand curves. And this approximation works rather well. This model might suffice over a decade or so in tracking movements of supply, demand, and oil imports.

Over a longer horizon, however, technological changes will shift the range of potential substitution in fundamental ways. Think only of the configuration of energy consumption at the turn of the century to realize that understanding technological change is the key to explaining differences between then and now. Careful analysis of turn-of-the-century technology in steam engine design or trade-offs between horse and horseless carriages are of little use today. Hence, when addressing policy questions that span decades, such as how to deal with greenhouse gases, there is no alternative but to adopt a view on trends in technology and the impacts on energy demand.

The work of Manne and Richels has been most prominent in the debate in highlighting the importance of technological change in the "autonomous energy-efficiency improvements" factor. In their model, technology is an exogenous trend that reduces energy consumption over time, other things being equal (Manne and Richels, 1992). The trend assumption is among the most important inputs, and there is very little guidance to help get an empirical fix on it. Worse, the rate of technological change cannot be truly exogenous for a large economy over a long period of time. Common sense dictates that scarcity, as summarized in relative prices, should substantially affect the rate of technology development. Thus, technological change is endogenous. It is affected by the path of economic development and energy prices.

Although this commonsense argument is easy to accept, it has had relatively little impact on energy modeling and analysis and even less on the communication of the analytical results intended to guide policymakers. Here, Lovins's more holistic approach is closer to the mark, but his approach is divorced from any formal model that exploits empirical evidence. The only formal modeling framework that attempts to capture both the theory of endogenous technological change and the empirical implications in estimation is that of Jorgenson and Fraumeni (1981: 17–47). Although their work and, later, that of Jorgenson and Wilcoxen (1993) have been extensively applied by their developers, they have not been widely emulated or fully utilized. Yet the available evidence suggests that many of the most popular assumptions about the nature and rate of technological change may be biased. For example, they may have the wrong sign (i.e., they assume positive technological change where a negative change may be correct) (Hogan and Jorgenson, 1991).

The theoretical foundations for modeling technological change are shaky, and empirical implementation is especially difficult. Analysis must concentrate on separating all the effects of substitution and the messy dynamics of energy and related prices. Once these other factors have been explained and controlled, the model of technological change tries to explain the residual factors that make up total factor productivity. Any hope of cracking the problem requires a general equilibrium analysis. The data and estimation requirements are daunting and have prevented most from following this path. It is surprising that more of the analyses and associated policy pronouncements have not incorporated the fruits of this research. This is a challenge for analysis and communication that lies ahead.

Consumer Myopia, Discount Rates, and Market Failure

An exciting area of energy analysis combines the old decision-analysis approach with a new twist to integrate the theory of options with the investigation of energy investment and conservation decisions. This line of work holds a key insight that may resolve an old conundrum regarding the disparity between consumer and corporate discount rates.

Consumers appear to substantially underinvest in cost-effective energy conservation technologies. The usual characterization of the problem in terms of consumer discount rates laments the myopia of consumers in failing to adopt cost-effective energy conservation technologies. That such overlooked technologies are in fact cost-effective usually derives from an engineering analysis of costs and benefits and a present-value calculation based on an electric utility's assumed cost of capital (approximately 12 percent). There are many difficulties in the cost analysis based on engineering projections (Joskow and Marron, 1992), but the self-evident proposition that consumer discount rates are too high is widely accepted, often based on the econometric work of Hausman, who found discount rates as high as 89 percent, depending on income level (Hausman, 1979: 53).

This interpretation of consumer choice as a market failure has contributed to the rationale for programs to substitute regulated utility decision making, discount rates, and choice of energy-efficiency investments. The rationale is that the utility makes supply-side investments in new generating facilities based on cost-effectiveness using its low cost of capital. A level playing field would require demand-side investment to be evaluated on the same basis. Of course, such cost-effective investments would require subsidies to overcome the consumer's unwillingness to invest alone. Resulting large-scale demand-side management (DSM) programs soon began to amount to a noticeable portion of total electricity costs.

The thrust of the new energy research is that this old analysis may have the story partly backwards (Jaffe and Stavins, 1993). The new insight draws from options theory and its application to energy decision making (Dixit, 1992: 107–32; Dixit and Pindyck, 1993). Here the argument harkens back to decision analysis and the value of information with sequential resolution of uncertainty. The usual cost-effectiveness analysis assumes certainty in estimates of investment costs and energy savings or, at most, performs a simple sensitivity analysis. But the analytical approach usually submerges the option value of waiting in the presence of uncertainty. If there is significant uncertainty about energy prices and technology, then the usual choice of investing in energy conservation is not a once-and-for-all decision, but one to invest now or to wait and revisit the issue when more information becomes available.

Investing now forecloses the option to wait. If the option has value, as it does when there is a significant chance that new information will change the decision, then the impact is to require a higher threshold of return before making an investment.

This higher threshold is indistinguishable from a higher discount rate applied to the existing "cost-effective" technologies. This recent research could go a long way toward explaining the otherwise "irrational" behavior of consumers.

This argument also cuts against utility decisions to invest in capital-intensive generating equipment. Investment in generating plant can also be compared to the option of waiting for more information. In the presence of significant uncertainty, this behavior would be indistinguishable from applying a higher discount rate to the evaluation of generation investment. Hence, the resolution of the conundrum is not that consumers are myopic and use discount rates that are too high, but that utilities are overconfident and use discount rates that are too low. The level-playing-field argument still applies, but its application is on a field of higher discount rates, smaller modular investments, and greater attention to flexibility in the face of uncertainty.

While it is not likely that this new perspective will resolve all the debate surrounding the proper balance between conservation investments and new generation, it shows how analysis and careful modeling can go a long way toward a powerful insight that has escaped many analysts not using the proper framework. More work remains to be done to pin down the magnitudes and communicate the ideas. The challenge lies ahead.

Insights Not Yet Acquired

There is a long list of important unsolved problems in energy analysis that inherit all the dilemmas of economic modeling in general. Perhaps the most persistent is cartel behavior. Early work on models of OPEC and its effects struggled with the difficulties of modeling anything other than the extreme cases of perfect competition and perfect monopoly. The range between these extremes covers too much and we know too little to be confident about anything we say in this domain, despite the importance of the questions.

Early published analyses of cartel behavior assumed that OPEC producers, or a subset, acted as a unified market leader surrounded by a competitive fringe (Pindyck, 1978). Originally it was thought that prices could not rise much above the level in 1974, as the present value-maximizing price had been reached or exceeded. However, additional analytical work not exposed until much later, but performed in 1975, revealed that the conclusion was entirely dependent on the seemingly innocuous assumption of linear demand curves. With linear demand, elasticities rise rapidly with rising price, soon reaching the economic ceiling for the monopolist. Changing to alternative demand variants that were equivalent over the range of the data but did not extrapolate rapid increases in elasticities produced a dramatically different result (Fromholzer, 1981; Wirl, 1985). This insight, accepted at the time, led to the conclusion that the government should not discuss the strong upside potential in OPEC prices. However, the practical result can be seen as the focus of the Federal Energy Administration's *National Energy Outlook* of 1976, which shifted to the higher price range compared with the *Project Independence Report* of 1974. The higher prices and revenues of 1979 and the early 1980s confirmed that a coordinated oil cartel could maintain higher prices and might wish to do so on behalf of its own interest.

Despite this one success, the problem for oil market analysts is that the cartel is anything but coherent and monolithic, and all formal attempts to get us further than the view of Adelman, summarized so well as the

"Clumsy Cartel," have been less than convincing (Adelman, 1980). Most advanced game-theory models, or extensive, diligent efforts to build simulation models to provide a better understanding of OPEC behavior, including my own, performed badly and reverted to the poorly supported target capacity utilization model that called for OPEC price increases when its production exceeded 80 percent of available capacity (Salant, 1982; Gately, 1977; Powell, 1990). Moreover, this default model was completely inconsistent with the high rates of capacity utilization and low oil prices of 1991–93. This model is useless, and there is nothing else on the horizon.

The contributions of formal modeling are limited to the valuable but narrow "under-the-lamppost" investigations where the light is better. For those who wish to examine the range of options that are available to the cartel, Gately's many sensitivity studies examining the interaction of prices and demand under a wide range of cartel pricing rules can be of real help (Gately, 1977). Many near-optimal pricing paths are available, and each pricing path is different. This will be of little comfort to OPEC, but it may in part explain why analysts with models have such a hard time making useful predictions in this domain.

Another perspective is the analysis of the competitive bounds on OPEC prices. A collection of papers from the time of the big oil price drop in 1986 demonstrates that this problem is amenable to analyses that point to a competitive price floor far lower than even the low oil prices of 1986 (Adelman, 1986; Griffin and Jones, 1986). As for the cartel, there is not much that can be said and believed based on the analysis of the formal models. The cartel is not dead, nor is it irrelevant. Prices are higher than the competitive floor and lower than the maximum that could be sustained. In between, we are all groping along with the clumsy cartel. Extensive and repeated analysis has not produced, much less communicated, any real insights about the behavior of the oil cartel or its effects on the world oil market.

Conclusions

Energy modeling and policy analysis is a fertile field for studying the interaction between ideas and the practical world. The surprise and impact of the 1973 oil embargo motivated an array of studies that still echo two decades later. Along the way, the practice of energy policy analysis changed with shifts in policy and gradual maturation of understanding of how to do and use formal policy analysis built on models and data. These illustrations highlight a prominent lesson concerning the interaction of analysis and policy, and of analysts and policymakers. The contribution of successful energy models has not been specific predictions or prescriptions, but insights into the complexities of the energy-economic system. Achieving success and improving the track record depend on a conscious effort to balance investment in analysis with investment in communication. In the marketplace of ideas, as in most markets, marketing is as important as production. Both marketing and production of analysis must adapt to be mutually reinforcing.

NOTES

The author is grateful to Carter Wall for her assistance in researching this chapter. The views presented here, and any errors, are solely the author's responsibility.

1. "And all the peoples came to Egypt to buy grain from Joseph, for the famine was grievous everywhere" (Gen. 41:57).

2. For those who doubt the disconnect between argument and analysis, there is succinct evidence in the 1988 exchange in *Energy Journal*, a presentation of three different views in what was billed as a debate (Tussing et al., 1988). Virtually any reader would conclude that the authors were talking past each other, not debating the merits of an analytical argument.

REFERENCES

Adelman, Morris A. 1980. "The Clumsy Cartel." *Energy Journal* 1 (1).

——. 1986. "The Competitive Floor to World Oil Prices." *Energy Journal* 7 (4).

Baumgartner, Thomas, and Atle Midttun. 1987. *The Politics of Energy Forecasting: A Comparative Study of Energy Forecasting in Western Europe and North America.* Oxford: Clarendon Press.

Dixit, Avinash K. 1992. "Investment and Hysteresis." *Journal of Economic Perspectives* 6 (1).

Dixit, Avinash K., and Robert S. Pindyck. 1993. *Investment under Uncertainty.* Princeton: Princeton University Press.

Doane, M. J., and Daniel F. Spulber. 1991. "Open Access and the Evolution of the U.S. Spot Market for Gas." Kellogg Discussion Paper 91-48. Northwestern University, Evanston, Ill.

Economic Report of the President. 1975. Washington, D.C.: U.S. Government Printing Office.

Energy Information. 1977. April 6.

Energy Modeling Forum. 1977. *Energy and the Economy.* Washington, D.C.: National Academy of Sciences Committee on Nuclear and Alternative Energy Strategies.

——. 1981. "Aggregate Elasticity of Energy Demand." *Energy Journal* 2 (2).

Fromholzer, D. 1981. *Demand Approximations for Oil Pricing Models.* Stanford, Calif.: Department of Economics, Stanford University.

Gately, D. 1977. "Strategies for OPEC's Pricing Decisions." *European Economic Review* 1.

Geoffrion, A. M. 1976. "The Purpose of Mathematical Programming Is Insight, Not Numbers." *Interfaces* 7 (1).

Greenberger, Martin. 1983. *Caught Unawares: The Energy Decade in Retrospect.* Cambridge, Mass.: Ballinger.

Greenberger, Martin., and William W. Hogan. 1987. "Energy-Policy Modeling in the US: Competing Societal Alternatives." In *The Politics of Energy Forecasting: A Comparative Study of Energy Forecasting in Western Europe and North America,* edited by Thomas Baumgartner and Atle Midttun. Oxford: Clarendon Press.

Griffin, J. M., and C. T. Jones. 1986. "Falling Oil Prices: Where Is the Floor?" *Energy Journal* 7 (4).

Hausman, J. 1979. "Individual Discount Rates and the Purchase and Utilization of Energy-Using Durables." *Bell Journal of Economics* 10 (1).

Havemann, J., and J. C. Phillips. 1973. "Energy Report/Independence Blueprint Weighs Various Options." *National Journal Reports* 6 (44).

Hogan, William W. 1979. "Dimensions of Energy Demand." In *Energy: The Next Twenty Years,* by Hans Landsberg et al. Cambridge, Mass.: Ballinger.

——. 1981. "Import Management and Oil Emergencies." In *Energy and Security,* edited by David Deese and Joseph Nye. Cambridge, Mass.: Ballinger.

——. 1983. "Oil Stockpiling: Help Thy Neighbor." *Energy Journal* 4 (3).

——. 1992. "Contract Networks for Electric Power Transmission." *Journal of Regulatory Economics* 4 (3).

——. 1993a. "A Competitive Electricity Market Model." Draft. John F. Kennedy School of Government, Harvard University, Cambridge.

——. 1993b. "Electric Transmission: A New Model for Old Principles." *Electricity Journal,* March.

——. 1993c. "Markets in Real Electric Networks Require Reactive Prices." *Energy Journal* 14 (3).

Hogan, William W., and Dale W. Jorgenson. 1991. "Productivity Trends and the Costs of Reducing CO_2 Emissions." *Energy Journal* 12 (1).

Hogan, William W., and A. Manne. 1977. "Energy-Economy Interactions: The Fable of the Elephant and the Rabbit?" In *Modeling Energy-Economy Interactions: Five Approaches,* edited by Charles J. Hitch. Washington, D.C.: Resources for the Future.

Holloway, Milton L., ed. 1980. *Texas National Energy Modeling Project: An Experience in Large-Scale Model Transfer and Evaluation.* New York: Academic Press.

Huntington, H., J. Sweeney, and J. Weyant. 1982. "Modeling for Insights, Not Numbers: The Experiences of the Energy Modeling Forum." *OMEGA: The International Journal of Management Science* 10 (5).

Jaffe, A. B., and R. N. Stavins. 1993. "The Energy Paradox and the Diffusion of Conservation Technology." Draft. John F. Kennedy School of Government, Harvard University, Cambridge, July 14.

Jorgenson, Dale W., and Barbara M. Fraumeni. 1981. "Relative Prices and Technical Change." In *Modeling and Measuring Natural Resource Substitution*, edited by Ernst R. Berndt and Barry C. Field. Cambridge: MIT Press.

Jorgenson, Dale W., and P. Wilcoxen. 1993. "Reducing U.S. Carbon Emissions: An Econometric General Equilibrium Assessment." *Resource and Energy Economics* 15 (March).

Joskow, P. L., and D. B. Marron. 1992. "What Does a Megawatt Really Cost? Evidence from Utility Conservation Programs." *Energy Journal* 13 (4).

Kalt, Joseph P. 1982. *The Economics and Politics of Oil Price Regulation*. Cambridge: MIT Press.

Kalt, Joseph P., and Frank C. Schuller, eds. 1987. *Drawing the Line on Natural Gas Regulation: Harvard Study on the Future of Natural Gas*. Westport, Conn.: Greenwood Press.

Leiby, P., and L. Russell. 1989. *Strategic Petroleum Reserve Planning Models: Numeric Comparisons*. Oak Ridge, Tenn.: Oak Ridge National Laboratory.

Lovins, Amory B. 1979. *Soft Energy Paths: Toward a Durable Peace*. New York: Harper Colophon Books.

Manne, A., and R. Richels. 1992. *Buying Greenhouse Insurance: The Economic Costs of Carbon Dioxide Emission Limits*. Cambridge: MIT Press.

Maryland People's Counsel. 1984. *Inquiry on Impact of Special Marketing Programs on Natural Gas Companies and Consumers*. FERC docket no. RM84-7-00.

Nelson, C. R., S. Peck, and R. G. Uhler. 1989. "The NERC Fan in Retrospect." *Energy Journal* 10 (2).

New York Times. 1974. "Project Manager Did It His Way," November 13.

Pierce, R. 1988. "Reconstituting the Natural Gas Industry from Wellhead to Burnertip." *Energy Law Journal* 9 (1).

———. 1993. "Update on Reconstituting the Natural Gas Industry." Draft.

Pindyck, Robert S. 1978. "The Gains to Producers from the Cartelization of Exhaustible Resources." *Review of Economics and Statistics* 60 (2).

Powell, S. 1990. "The Target Capacity Utilization Model and the Dynamics of the World Oil Market." *Energy Journal* 11 (1).

Public Papers of the Presidents, September 22, 1975.

Salant, Stephen W. 1982. *Imperfect Competition in the World Oil Market*. Lexington, Mass.: D. C. Heath.

Stanfield, R. 1985. "Supporters of the Strategic Petroleum Reserve Say 'Fill It to the Brim.'" *National Journal*, May 11.

Teisberg, T. 1981. "A Dynamic Programming Model of the US Strategic Petroleum Reserve." *Bell Journal of Economics* 12 (2).

Tussing, A. R., et al. 1988. "Is an Oil Tariff Justified? An American Debate": A. R.

Tussing and S. A. Van Vactor, "Reality Says No"; H. G. Broadman and W. W. Hogan, "The Numbers Say Yes"; D. M. Nesbitt and T. Y. Choi, "The Numbers Say No." *Energy Journal* 9 (3).

U.S. Department of Energy (U.S. DOE). 1987. *Energy Security.* Washington, D.C.

U.S. Department of Energy, Energy Information Administration (U.S. DOE/EIA). 1992. *Monthly Energy Review.* Washington, D.C., February.

U.S. Federal Energy Administration (U.S. FEA). 1974. *Project Independence Report.* Washington, D.C.

——. 1976. *National Energy Outlook.* Washington, D.C., February.

U.S. General Accounting Office (U.S. GAO). 1983. *Status of the Strategic Petroleum Reserve as of December 31, 1982.* GAO-RCED-83-93. Washington, D.C.

Verleger, Philip K. 1982. *Oil Markets in Turmoil.* Cambridge, Mass.: Ballinger.

Weimer, David Leo. 1982. *The Strategic Petroleum Reserve: Planning, Implementation, and Analysis.* Westport, Conn.: Greenwood Press.

Wirl, F. 1985. "Stable and Volatile Prices: An Explanation by Dynamic Demand." *Optimal Control Theory and Economic Analysis* 2.

8. Energy Modeling and Energy Policy

MICHAEL E. CANES

William Hogan's basic thesis (chap. 7), in brief, is that strong analytic insights coupled with good communication of those insights to policy-makers yield good policies. There is much of value in Hogan's argument, but I disagree with the basic thesis and will offer an alternative explanation of the link between analysis and policy.

Does Good Communication Lead to Good Policy?

Hogan argues that insights gained from the process of energy modeling are more important than results obtained. Understanding the interactions among variables of interest is valuable, while particular forecasts or predictions are less so. And if the forecasts run counter to what an economist's intuition tells him, they are probably wrong. These are good points. Intuition is probably sufficient if all we want to know is whether a given policy will move us toward a given goal. However, understanding the interrelationships among the variables of interest and between those variables and others requires more than casual evaluation.

Hogan's emphasis is not on the mode of analysis so much as it is on good communication of analytic insights. While it is difficult to argue with the idea that good communication helps to promote better thinking by policy-makers, this emphasis seems to me misplaced.

Hogan provides four examples of insights acquired and communicated (i.e., policy successes): application of economic analysis to the notion of energy independence at the time Project Independence was proposed (1974); estimation of supply and demand elasticities in energy markets and their implications for the relationship between energy use and economic growth;

analyses of the effects of price and allocation controls on energy markets; and analyses of competitive forces within natural gas and electricity transportation markets despite the apparent presence of natural monopoly.

These examples basically ring true. The thinking at the time of Project Independence was based on a supply-demand "gap," wherein domestic energy supply fell short of demand. It was thought that strong government action was needed to fill the gap. Economic analysis revealed that supply and demand would balance at a price that would allow energy markets to function appropriately and that any program to achieve U.S. energy independence would be impossibly expensive. As a consequence, the notion was quietly dropped. Score one for analysis.

Hogan also argues that estimates of energy demand and substitution elasticities led to the conclusion that there was no need to subsidize energy markets to maintain or encourage economic growth. Empirical analysis showed that other resources are sufficiently substitutable for energy that one need not treat energy as a special growth input. I agree with this conclusion, but Hogan takes it too far.

Specifically, he argues that demonstration of relatively large energy demand elasticities means that economic growth could proceed almost indefinitely without corresponding growth in energy use. Energy-saving capital and technology would provide the difference. This is true, *if* one is willing to apply capital and technological resources at ever higher rates to energy saving. But why do that? Surely the same analytic insight that reveals that growth can proceed with reduced energy use also reveals that constraining energy use imposes constraints on growth. Data relating changes in growth and energy use in Western industrialized countries after the oil price increases of 1973 and 1979 support this assertion (e.g., see Maddison, 1987).

Finally, Hogan cites analytic work that led policymakers to conclude that natural gas and electricity markets could be opened up to competition even though there might be but a single pipeline or electricity grid in any given geographic area. The basic argument was that sellers of the raw good (e.g., gas or electricity) could compete to sell to ultimate buyers if they could be assured access to the pipeline or grid at terms no worse than the pipeline or electric company offered its own sales subsidiaries. I agree that this is a policy success, and analysis clearly has helped fashion how to price the relevant services, properly separate them, and make the transition from the previous one-seller world to a world of competing entities. From my perspective, however, the mystery is that the notion of competing sellers required such sophisticated analysis in the first place. After all, we have had common carriage in the interstate oil pipeline business for a very long time, so there was a working model available to policymakers from the beginning. While oil can move economically by means other than pipeline in some markets, the basic physical principle of many sellers over a single transportation conduit was evident.

Hogan cites two examples of policies he considers policy failures: inability to use the Strategic Petroleum Reserve (SPR) even when oil market conditions in 1990 clearly indicated its utility; and unwillingness to affix some sort of fee on imported oil even though (some) analysts have concluded that it imposes (correctable) external costs on society. In both cases, Hogan believes that the problem has been not a lack of good analysis, but a failure to communicate adequately the results of analytic insights. Analyses exist, he acknowledges, that suggest that an oil import fee might not be desirable. But he believes that these are essentially based on ideological notions of free trade, independent of the particular conditions of the oil market.

This is not the place to engage in a policy debate on oil import fees. However, I believe Hogan gives rather short shrift to analysts who reach different conclusions from his, and he also understates the practical problems with such a policy. The work of Bohi and Toman, for example, suggests that much of what are sometimes called externalities associated with oil imports are not externalities, or at least have not been proven to be so (Bohi and Toman, 1993). Further, the United States now has free-trade agreements with Canada and Mexico, so that the institutional structure surrounding an oil import fee would have to police transshipments from third countries through those two trading partners.[1] Beyond that, there are problems with the political economy of an oil import fee. The ideal version that analysts have in mind is nothing like what the political process would in fact fashion. Hogan dismisses this issue, but experience with the oil import quota program in the 1960s, and with the oil entitlements program in the 1970s, persuades me to give it considerable credence.

Hogan provides two examples of insights acquired but not communicated. The first deals with the relationship between technological change and energy demand, where the work of Jorgenson and various colleagues is cited as trying to determine what causes technological change. In a co-authored article, Hogan makes the point that, contrary to popular opinion, a constraint on energy demand—imposed, say, by an energy tax—would reduce, not enhance, technological change (Hogan and Jorgenson, 1991).

A second example has to do with what explains so-called consumer myopia with regard to energy-saving investment. Here Hogan develops the idea of a sort of *option demand,* whereby consumers always have the option of making an energy-saving investment now or waiting until later. The resulting behavior is identical to what would happen with unrealistically high discount rates, but it has a rational economic motive behind it. Both of these seem to me to be good examples of insights acquired but not yet fully communicated. I agree with Hogan that therein lie future opportunities.

Finally, Hogan cites understanding of the behavior of OPEC as a problem yet to be conquered by analysts, particularly in the light of what has happened over the past few years. While we have some insight into parameters of the world oil market, I agree that much of what passes for OPEC analysis is ad hoc, made to fit the facts of the moment. Whether further analytic work is a future opportunity or a blind alley remains to be determined. A lot of effort has gone into this, and we have made surprisingly little progress so far.

Communication Isn't the Problem

My criticisms of Hogan's article are threefold: minor quibbles; concerns regarding the logic of his thesis; and questions about how he defines *success, failure,* and *communication.*

First, a quibble. Hogan describes his figure 7.1, a graph of OECD gasoline demand versus price, as being "as close to the textbook example of a highly elastic demand curve as we are ever likely to find." From appearances, however, it looks elastic over one range, perhaps unitary elastic over another, and quite inelastic over the relatively higher price range. Evidently elasticity is in the eye of the beholder.

On logic, there is a problem with Hogan's thesis that good analytic insight plus good communication of that insight leads to good policy. According to that thesis, if a particular policy is successfully implemented, then it must have good analysis underlying it and was satisfactorily communicated. But if it is not successfully implemented, there must have been inadequate analysis or poor communication (generally the latter, in Hogan's view). The trouble is, this leaves us no useful insight into how to achieve better policy, since we have no criterion other than its ultimate success as to what is sufficiently good analysis or communication.

Further, in the two cases Hogan cites as policy failures, I disagree that there has been inadequate communication. As far as use of the SPR goes, energy analysts have done quite a bit of work on how it should be used, and even on mechanisms to encourage its proper use (see, e.g., Bohi and Toman, 1987a and 1987b). And in this particular instance a powerful politician in a position to influence policy—Congressman Phil Sharp, former chairman of the Subcommittee on Energy and Power of the House Subcommittee on Energy and Commerce—has known about and understood the policy quite well.

Rather, the problem is more likely that political criteria take precedence over economic ones. Since release of oil from the SPR is a political decision, policymakers will consider the political signals from such a release. These signals range from public perception of the crisis at hand to how other governments might react. For example, the U.S. government may not want the public to perceive a given situation as an emergency, and it may want to be sure other governments do not react in a manner that would offset a prospective SPR release.[2]

In addition, the incentive structure facing decision makers regarding when to release oil from the SPR may be biased, stemming from how they likely will perceive the consequences of acting too early as opposed to too late. If they act too late, they can be accused of not preventing as much economic damage as they might have—exactly the criticism Hogan levies. This would risk criticism by a few energy economists. But if decision makers act too early, it will be apparent to the entire body politic that the SPR has been drawn down too far, in advance of real trouble. What defense would they then have against a charge of imprudence—that they were trying to limit economic damage and did not consider it likely that things would get considerably worse? That is not likely to assuage many voters. Thus, excessive prudence looks like the policy choice in most circumstances.

As for the question of an oil import fee, I believe Hogan is far too modest about his own efforts to communicate on this subject. Having read a number of his papers on the matter, I would bet he has offered his views before a wide range of policy-oriented audiences. Nor is he alone in this communication. Other analysts both here and abroad have proposed such fees. I do not believe that inadequate communication is the problem with this proposal. Rather, there are respectable analyses that reach quite different conclusions. There also are several practical problems, and the political calculus of a significant oil import fee suggests that it is not a likely policy choice.

Finally, I question Hogan's use of the terms *policy success* and *policy failure*. Hogan believes that analysis led to the demise of the notion of energy independence, but realistically, how do we explain the ever-expanding sub-sidies being given to alternative motor fuels by federal, state, and local governments? Isn't the goal of substantially reducing dependence on for-eign oil driving these subsidies? And aren't mandatory energy conservation standards also driven in part by such thinking?

Similarly, Hogan states that good analysis led to abandonment of the notion that energy subsidies are needed for economic growth. Maybe so, but it cannot escape notice that one of the strongest arguments given for subsidies to renewables, electric vehicles, energy conservation, and the like is that development of such technologies will stimulate U.S. global eco-nomic leadership, whole new industries, and job and productivity growth.

What about ending price and allocation controls on oil and gas? This was a good policy outcome, but it took ten years to achieve in the case of oil (1971–81), and almost forty years in the case of gas (1954–93, when the last of the controls were lifted). With policy successes like that, we don't need many failures.

The natural monopoly/competition model for gas pipelines and the elec-tric grid might be the best example of an unambiguous policy success, but as stated earlier, a working model was available all along, and the wonder is that it took so much analytic power to transfer its findings to these markets.

What about the meaning of *failure?* While one could argue that the SPR has not been used the way it should have been, we must recognize that policymakers responded to what analysts were recommending by sinking over $20 billion into the enterprise—hardly an insignificant policy choice. As for an oil import fee, there is in fact a small one on imported products. Moreover, the Comprehensive Environmental Response, Compensation, and Liability Act (CERCLA, or "Superfund") also imposes a small fee. Still, Hogan is correct that the policy has not been adopted, but as I have indi-cated, the analytic world is far from united on this subject.

An Alternative Thesis

Hogan's data regarding energy policy successes and failures can be better explained by a different thesis. By and large, where economic analysis has concluded that energy markets would better direct resources than govern-ment, that analysis has led to policy success. Conversely, where economic analysis has concluded that government resource direction would outper-form markets, the policy has failed. This thesis reasonably explains why all of the six policies Hogan reviews come out as successes or failures, although I concede that creation of competitive market conditions in gas pipelines and electric grids has taken some very active institutional restructuring by pertinent government agencies.

Why should market-oriented policy analysis work out better than interventionist policy analysis? I conjecture that economists understand market behavior very well, and therefore are able to provide policymakers real insight into how markets can resolve difficult resource problems. On the other hand, economists have very limited understanding of political behavior, and therefore their expectations regarding the outcomes of government intervention into markets are often wrong. The earlier example of the SPR, which has been built but not used when needed, is instructive.

All of this leads to two final comments. First, there is much for energy policy analysts to do, as there are several instances where current policies raise questions regarding efficient resource use. Examples include expanding subsidy programs for alternative fuels of all kinds, and technology forcing through, for example, electric vehicle mandates. Energy analysts also might seek to repair flawed policies (e.g., by finding politically acceptable means to relieve public-sector officials of the burden of choosing when to use the SPR). In this regard there are future opportunities.

Second, energy policy successes will come more readily from analyzing how energy markets could or do work to allocate resources, than from focusing on the communication of results. It is not that communication does not have a role; it is just that economists have never been shy about publicizing their findings. With communication of information becoming cheaper, it is hard to worry about how much of it there is.

NOTES

1. Sweeney (1989) argues that a fee with exemptions such as these would result in net welfare losses to the United States.

2. During the Persian Gulf War, for example, the United States would not have wanted to do anything to reduce Saudi incentive to increase production as fast as possible.

REFERENCES

Bohi, Douglas R., and Michael A. Toman. 1987a. "Futures Trading and Oil Market Conditions." *Journal of Futures Markets,* April, pp. 203–21.

——. 1987b. "Restructuring the IEA Crisis-Management Program to Better Serve

Member Country Interests." In *Responding to International Oil Crises,* edited by George Horwich and David L. Weimer. Washington, D.C.: American Enterprise Institute.

———. 1993. "Energy Security: Externalities and Policies." *Energy Policy,* November, pp. 1093–1109.

Hogan, William W., and Dale Jorgenson. 1991. "Productivity Trends and the Costs of Reducing CO_2 Emissions." *Energy Journal* 12 (1): 67–85.

Maddison, Angus. 1987. "Growth and Slowdown in Advanced Capitalistic Economies: Techniques of Quantitative Assessment." *Journal of Economic Literature* 25 (2): 649–98.

Sweeney, James L. 1989. "Oil Import Fees with Exemptions: An Empirical Examination." *Resources and Energy* 11:215–39.

9. Past and Present Failures as Guides to the Future

GLENN R. SCHLEEDE

The period surrounding the energy shock of 1973–74 was rife with attempts by the federal government to define and implement national energy policies. We can learn a lot from the period since the shock if we analyze failed federal policies and programs as well as successes. As usual, the benefits of hindsight are helpful. In sum:

—Improvements in the nation's energy situation during the past twenty years were more often due to market and private-sector efforts than government policies.

—The policy analysis community, within and outside the federal government, is eager to take credit for success but reluctant to identify and acknowledge failure, or to address the tough issues that remain from the 1970s and 1980s. While optimistic about the nation's energy future, I am pessimistic about the potential for a constructive role for government in most energy matters.

Hogan and Modeling

William Hogan's analysis (chap. 7) is limited to lessons that can be learned from energy modeling and analysis in the two decades since the 1973 oil embargo. The scope of his analysis is far too limited to derive insights from the policy and program successes and failures of the period. Within the area treated by Hogan the analysis is commendable for recognizing the limitations of econometric modeling and explicitly acknowledging that models should be used to gain insights, not to make forecasts. Students and public policy analysts would be well served if they heeded his advice.

The Continuing Forecasting Problem

However, widespread use of econometric models in energy market forecasting, spawned during the 1970s, cannot be dismissed merely by noting that the *numbers* produced by forecasts are generally incorrect, or that they should not be used as the basis for decisions. The U.S. Department of Energy (DOE) regularly produces long-term forecast numbers that are used to justify policy proposals and positions. The Energy Information Administration (EIA) publishes an *Annual Energy Outlook* with supply, demand, and price forecast numbers extending through 2010. These are used as the basis for decisions within, and in some cases outside, government.

Also, some organizations outside the federal government have made millions of dollars by producing energy forecasts with specific numbers developed from econometric models. These forecasts continue to be used as the basis for real-world decisions. They are regularly used to evaluate a variety of potential long-term contracts and investments in facilities to produce, transport, use, or conserve energy. Even executives who profess not to believe *any* forecast knowingly or unknowingly rely on someone's energy market forecast whenever they rely on an economic analysis of the net present value or internal rate of return of potential long-term investments and contract commitments.

Billions of consumer, shareholder, and taxpayer dollars have been wasted because investment and long-term contract decisions by energy company executives and regulators were based on forecasts that turned out to be incorrect. The role of forecasts in real-world decisions cannot be dismissed. Forecasters and the energy analysis community have a responsibility to deal with this continuing problem. Minimally, they need to make clear that their forecasts are driven by assumptions that may or may not be correct; that different assumptions and models are likely to yield very different forecasts; and that potential decisions should be tested against forecasts that reflect differing perceptions of future energy market circumstances.

Incorporating Technological Developments into Models

Another valuable suggestion of Hogan's is that analysts need to find a way to incorporate technological developments into energy market models. Changes in technology due to market forces, regulatory requirements, electric utility subsidies for demand-side measures, or "spinoffs" (e.g., more efficient gas turbines from aircraft engine R&D), will continue to have a major impact on energy markets. Also, developments in computer, information, communications, electronic controls, and materials technologies that are likely to improve energy efficiency and to reduce energy-intensity should be taken into account in energy market models.

Changes in technology will likely affect the U.S. industrial mix and industrial sector energy-intensity. While it is difficult for analysts to estimate the impact of technology developments, Hogan's recommendation is worth pursuing. However, the effects of these developments on energy demand are unlikely to await a full decade, as Hogan suggests.

Defining Policy Analysis

Unfortunately, Hogan seems to consider *economic analysis* as synonymous with *policy analysis*. Economists have an important role to play in energy and environmental policy analysis, but bright people with backgrounds in a variety of academic disciplines will continue to make important contributions to the analysis of energy policies and programs. Nearly all of the important policy issues that must be addressed will have to be approached on an interdisciplinary basis that includes contributions from the hard sciences, engineering, social sciences, and other fields.

Furthermore, Hogan identifies only a few of the proposals advanced during the early 1970s. As demonstrated below, many others were made, but only a few were successful. Many were failures or were not adopted. Moreover, there are good reasons to question whether economic modelers should be credited with the role Hogan suggests, let alone with the success in changing policies following the energy shock. For example, problems with price controls were well known before energy modelers arrived on the scene.[1] Also, the initial Nixon administration proposal to deregulate natural gas *preceded* the oil shock. Credit and blame for success and failure should be assigned to the longstanding approach to policy formulation known as "muddling through."

Disagreements versus Not Communicating

Hogan identifies "emergency management and the Strategic Petroleum Reserve" and "energy security and oil import fees" as examples of "insights acquired but not communicated." He confuses disagreement with poor communication. Many of us are tempted to conclude that failure of others to accept our conclusions is due to a failure to communicate (often with the implication that the party rejecting our wisdom is not listening to our analysis or is not very bright). More often, rejection is due to others disagreeing with our conclusions. At times they are correct.

Hogan's examples represent substantive disagreement with the analysts' conclusions rather than a failure to communicate. Apart from the weight given to the benefits of trade and low-cost energy supplies, many who *reject* proposals for oil import fees legitimately disagree with the assumptions underlying the analysis proffered by import fee advocates. They are also concerned about the micro and macro effects of proposed fees. Finally, those opposing fees also have legitimate concerns about the potential for inefficient use of the tax revenues.

In the case of arguments over the establishment of the Strategic Petroleum Reserve (SPR), the Office of Management and Budget has a continuing obligation to the presidency to present analyses and arguments that will help the sitting president in his nearly impossible task of allocating short resources among many competing demands in the president's annual budget. Further, those clamoring for a commitment to withdraw crude oil from the SPR during, for example, the Persian Gulf War miss one of the SPR's greatest values: having an SPR available to use provides a good excuse for not accepting the spate of unwise, uneconomic, and counterproductive proposals that are invariably offered by Congress and others when energy prices begin to rise or a fear of shortages emerges.

Nixon's "Energy Independence"

Hogan correctly concludes that the objective of "energy independence," if interpreted as energy self-sufficiency, made no sense in 1973–74 and makes no sense today. Use of the phrase, which was not uniformly endorsed within the White House at the time it was used, is partly attributable to two factors Hogan did not mention:

—Naiveté on the part of presidential speechwriters, who are too often successful in substituting high-sounding rhetoric for feasible, commonsense commitments.

—A faulty perception left over from President John F. Kennedy's "commitment" that the United States would have a man on the moon before 1970. The success in meeting that unique goal (at great cost) has led many to conclude that the federal government can achieve *any* goal—including the elimination of crime or the drug problem—if a president announces the goal and throws enough money at it. Fundamental differences in goals do not deter some advocates from using faulty analogies.

Nixon and Ford Energy Proposals

A fair evaluation of lessons requires a more comprehensive approach than that taken by Hogan. To that end I reviewed major energy statements, messages, and fact sheets emanating from the White House and the Executive Office of the President on thirty-one occasions between 1971 and 1976 when Presidents Richard Nixon and Gerald Ford were in office.[2] Appendix A provides a matrix showing actions announced and proposals made, by date. More than 150 actions and proposals are included. Some that required congressional action were made repeatedly. For example, removing natural gas wellhead price controls was mentioned ten times. The proposals cited in Appendix A fall into thirteen categories: research, development, and demonstration efforts (RD&D); leasing federal lands; energy production and transportation; environmental protection; oil storage and emergency reserve supplies; energy efficiency and conservation; fuel switching; energy allocation and price controls; taxes and fees; international activities; state public utility commission actions; organizational changes; and miscellaneous actions.

The Effectiveness of the Nixon and Ford Energy Proposals

A thorough evaluation of the success or failure of the Nixon and Ford proposals and actions is beyond the scope of this chapter, but it should be clear to the nation's energy policy analysts and historians that they have a fertile area for study. I offer the following evaluation based on my experience in government during and after the energy shock, and in energy market developments since the mid-1970s.

First, many of the proposals were unsupported by analysis and were not susceptible to the type of modeling and analysis Hogan describes. Little information was available that could be used in analyses. Questions raised about proposals by analysts in the Office of Management and Budget and the Council of Economic Advisers were often brushed aside. Proposals were often hastily adopted, purely on the basis of preferences and perceptions of persons in key energy positions, or were put forth by special interests and members of Congress. Many of the proposals, particularly those pertaining to RD&D, had been "sitting on the shelf" in agencies, awaiting an opportunity for funding, or were hastily developed in the crisis atmosphere to take advantage of free-flowing federal tax money.

Second, proposals for continued or increased funding for RD&D programs dominated the energy spending agenda in both the Nixon and Ford administrations, even though no one claimed near-term contributions from these programs.

Third, despite billions of dollars in federal spending and tax breaks for energy RD&D, federally subsidized programs have produced little in the way of technologies that can compete in the marketplace. This has not deterred the flow of subsidies.

Fourth, oil allocation and price controls failed and made matters worse as documented by Hogan, by Doug Bohi and Joel Darmstadter (chap. 2), and by others.

Fifth, energy markets were badly distorted by mistaken assumptions about U.S. natural gas reserves. Many have now noted the fallacy of preserving natural gas for "higher-value" use—begun in the Nixon administration and continued during the Ford and Carter administrations.

Sixth, reorganization of federal functions played a prominent role in presidential energy proposals. The prominence of such proposals reflected many factors, including the fact that *energy* had not previously been viewed as a crucial reason to consolidate functions. In addition, reorganization proposals were partly the result of internal power struggles; a need to show that something was being done; a search for spokespersons who could exert a calming influence in a crisis atmosphere; a desire by some to expand the role of the Atomic Energy Commission and its national laboratories to encompass all forms of energy; a recognition that the government had little hard energy data; and the lack of a "home" for efforts to encourage conservation.

Seventh, the actions contributing the most positive effects in the short and mid-term were those stimulated by higher energy prices. Actions that permitted use of coal in lieu of oil in electric generating plants and exhortations to conserve energy also helped in the short term. Other actions that contributed in the mid-term were those that accelerated leasing of outer continental shelf lands, cleared the way for construction of the trans-Alaska oil pipeline, and, most important, removed oil allocation and price controls and controls on wellhead natural gas prices. The largest reductions in oil consumption in the mid-term resulted from the construction of coal-fired generating units, completion of nuclear power plants that were already committed to be built, and improvements in automobile fuel economy.

Eighth, many administration proposals were not approved by Congress, and many unilateral actions taken by the administration were short-lived.

Ninth, the "crisis atmosphere" that prevails in Congress and the administration when there is a crisis or perceived crisis, such as energy shortages or high prices, does not contribute to rational policy development. The perceived need to "do something even if it is wrong" overwhelms good judgment. In fact, taking no action is often more beneficial than adopting unwise policy.

Tenth, government agencies—in this case, energy agencies—once started, are very difficult to trim back or eliminate, even when the initial reason they were created has disappeared.

Progress: Reliable Supply at Reasonable Cost and Minimal Environmental Impact

The public, the media, and politicians have been less concerned about energy since late 1980, except during the Persian Gulf War, during the 1993 threat of a broad-based energy tax, and as a result of occasional interruptions of electric power. Concerns are confined largely to environmental agencies and advocacy groups, government energy agencies, and energy industries.

Thankfully, we seem to have gone beyond a time when a speaker could evoke cheers from almost any energy industry audience by declaring, "We need a national energy policy!" There is now broader recognition that those using this phrase had widely differing views concerning the problem to be addressed and the policy needed. When pressed, it became clear that the policy desired was one that would benefit the speaker's organization or constituency—often to the detriment of others.

Nevertheless, energy is extremely important to our national welfare, and concern about the nation's energy outlook will continue to be a legitimate subject for attention and analysis. Market conditions reveal reasons for both considerable optimism and some pessimism:

—Proved world oil reserves are now 50 percent larger than in 1973, and proved world natural gas reserves are roughly four times those of 1973. Reserves are not evenly distributed, but many of the countries where reserves are concentrated need hard currency, and export of oil and gas is a way to acquire it. This is one reason for less concern about growing U.S. dependence on imported oil. Other developments—including the end of the Cold War, successful oil exploration in areas outside OPEC, the potential for increased oil production from the former Soviet Union (if Western technology is applied), and increased substitution of natural gas for oil in electric generation in various countries—have contributed to reduced concern about "vulnerability" to prolonged interruptions of world oil supplies and/or sustained high prices.

—Removal of oil and gas price regulation has resulted in lower, not higher, prices. Actions under way to reduce regulation and increase competition in the electric and natural gas industries hold additional promise for consumers.

—U.S. energy users have ready access to plentiful supplies, and there is vigorous and growing interfuel competition among oil, coal, and natural gas that helps hold down prices.

—*Real* energy prices are well below the high point reached in 1980–81, and in some cases below pre-energy-shock prices.

—The United States has avoided most attempts to impose energy taxes, except in motor fuels, thus avoiding many of the uneven tax burdens that fall heavily on families, organizations, industries, and regions most dependent on energy. Avoiding energy taxes has helped preserve an advantage for U.S. energy-dependent industries in global competition.

—Energy efficiency has continued to improve without resorting to taxation, though some improvements have been achieved by regulation and subsidized electric utility "demand-side" programs.

—Improved technologies developed in the private sector for producing, converting, and using energy are contributing to declining real prices and increasing efficiency and productivity.

—Improvements in computer, communications, information, electronic control, and materials technologies are very likely to contribute to additional improvements in energy efficiency.

—The need to protect the environment while producing, transporting, converting, and using energy is now widely recognized and increasingly incorporated into energy decisions.

—The collection and analysis of historical data on energy supply, demand, and prices have contributed in a major way to the nation's understanding of its energy situation and outlook.

Opportunities and Lessons Learned

Although the nation's energy outlook has improved substantially since the 1973–74 and 1979–80 shocks, many challenging opportunities remain for courageous public policy analysts. These are not necessarily new opportunities. Many of the issues are left over from the 1970s and 1980s, suggesting either that they are intractable, that the federal government is avoiding them, or that the policy analysis community is reluctant to address them. An objective analysis of policy failures could make a significant addition to the list of lessons to be learned from actions taken during and after the energy shocks.

However, effective analysis alone will not assure policy changes. Many of the issues involve the interests of influential constituencies that benefit from subsidies. They also affect entrenched bureaucracies, powerful members and committees of Congress, and the political philosophies of executive branch officials.

The following are examples of policies and programs that need analysis and evaluation but are being ignored.

Taxpayer-Subsidized RD&D

Perhaps the most glaring example of policies that have often failed but continue to be applied is the expenditure of billions of tax dollars to subsidize energy RD&D efforts. Despite expenditures of over $100 billion since the early 1950s, the U.S. government has little to show for these tax dollars in the way of technologies that can compete in the private sector. Successful contributions, such as improved gas turbines used in electric generation, are principally spinoffs from defense-related R&D.

Furthermore, to this day DOE lacks clear criteria for selecting and evaluating energy RD&D efforts, clear standards for information that must be presented by advocates, an effective process for establishing priorities, procedures to assess objectively the potential marketability of technologies if they are developed, or effective measures to assure that federal subsidies do not displace private investments in RD&D. Protestations by advocates of federal energy RD&D subsidies who argue that private companies also spend money on unproductive RD&D efforts carry no weight. When private companies find that efforts are unproductive, they cut off spending, an action that DOE and Congress avoid.

DOE Laboratories

Several studies have shown that DOE laboratories have in many cases run out of missions, but DOE has not been able to cut them back significantly or close them. Instead, DOE continues to funnel billions to the laboratories, often for low-priority work and without seriously considering whether other organizations could perform the tasks equally well at lower cost. Spending has been increased for attempts to find new, private-sector-oriented missions for the laboratories even though ample experience indicates that it is not practicable to redirect *organizations* that have grown up under government cost-type contracts to new private-sector missions. These organizations have little knowledge or experience in markets, and their costs of doing business tend to make them uncompetitive.

Furthermore, little thought has been given to the benefits of cutting back laboratory funding and allowing the dispersal of their highly educated professional staffs to private industry, colleges, universities, and high schools. This talent could provide an opportunity for such organizations to rebuild

146 UNRESOLVED ISSUES AND THE ROLE OF POLICY ANALYSIS

their talent banks following years of difficulty in competing with the salaries paid to scientists, engineers, mathematicians, and other professionals by government cost-type contractors.

Pilot Testing and Effective Program Evaluation

Federal agencies often launch major new efforts without pilot testing or a clearly defined evaluation program. For example, millions of tax dollars are being spent on cooperative research and development agreements (CRADAs), awarded with minimal review, on the theory that organizations developed for specialized work and nurtured on cost-type contracts can contribute significantly to the private sector. Obviously, these taxpayer-subsidized awards are popular with beneficiaries, who will attest to their great value. However, objective evaluation and assessment of CRADAs might yield a different answer. Evaluation is overdue. CRADAs are but one example. Dozens of questionable energy policies and programs adopted during the past twenty years could, if analyzed objectively, provide lessons for the future.

Industrial Policy

The Bush and Clinton administrations have had strong advocates for a federal role in advancing technologies intended largely for private-sector use. Industrial policy advocates seem undeterred by past failures, such as the federal effort to create a large, sustainable civilian nuclear power industry—one of the largest industrial policy experiments of all time. Would nuclear waste management, safety, and proliferation issues have been addressed in a more timely and effective manner if the industry had developed outside the direct sponsorship and protection of the Atomic Energy Commission and Joint Congressional Committee on Atomic Energy? Views on industrial policy are often ideological. Thus, good analysis may have no impact on those who believe that the federal government should play a leading role in picking and promoting development of particular technologies intended for the private sector, rather than allowing market forces to determine whether and when technologies are developed. Nonetheless, objective evaluation of industrial policy experiments is needed.

Balancing Energy and Environmental Objectives

Careful planning of energy projects and actions to reduce adverse environmental impact can increase the compatibility of energy and environmental objectives. However, some conflict between these objectives, even if minor, is often present. The U.S. Environmental Protection Agency (EPA) and other agencies with environmental responsibilities have the advantage of extensive resources to analyze energy-environmental conflict issues and to marshal data and arguments. DOE has devoted fewer resources to such efforts and has been less well equipped to deal with issues involving conflicting objectives. A reasonable balance is more likely to be achieved, and the public interest better served, if issues involving conflicting goals are approached through an integrated use of science, economics, and policy. There needs to be a balance in the abilities of agencies defending their assigned objectives. DOE should anticipate conflicts between energy and environmental objectives and then undertake or support analysis of the issues so that it can play a knowledgeable role. It should be as well informed about energy and economic impacts as EPA is about environmental impacts.

Establishing Environmental Priorities

It is now widely recognized that the federal government has not done well in taking cost, risk, and benefits into account when setting environmental requirements, or in establishing priorities among environmental objectives. In some cases EPA is prohibited by law from taking cost-benefit considerations into account in setting standards. Competition for both public and private resources will, in the years ahead, almost certainly require greater attention to prioritizing objectives and to cost-benefit considerations. The lead responsibilities will remain with EPA and other agencies with primary responsibility for implementing environmental statutes and regulations. However, DOE should get prepared to participate in these debates.

Tax Subsidies

Tax subsidies have long been used to promote development of energy resources, but with mixed results. They were expanded in the 1970s and early 1980s, in some cases to encourage development of various technologies, including wind power and solar energy. Subsidies were reduced in 1986, but some remain. The Energy Policy Act of 1992 added a production tax credit for electricity from generating units using certain renewable energy sources. Energy tax subsidies deserve continuing attention because they often distort energy markets. One deserving after-the-fact evaluation is the now-discontinued "Section 29" tax credit, which provided some $7.5 billion in benefits for producers of methane recovered from coal seams and tight gas formations. At one point the value of this credit closely approached the market price at the wellhead for natural gas, undoubtedly distorting markets.

Another tax credit worthy of attention is that for ethanol, which at the federal level alone exceeds the cost of crude oil. Considerable analysis has already criticized this tax credit, but the political clout of its principal producer and other advocates has prevented change. William Hogan's emphasis on persistent analysis, even in the face of political opposition, applies to this case.

Defining an Appropriate Federal Role

Federal government involvement in energy matters grew dramatically following the energy shock. Since then the regulatory role has diminished in some cases (e.g., oil and natural gas) and increased in others (e.g., energy-efficiency standards). The information role has increased, as has the promotional role in energy efficiency and conservation and attempts to dictate "national energy policy." Subsidies for energy RD&D have grown. The bureaucracy, including DOE and its many contractors and consultants, has also grown.[3]

Considerable disagreement remains as to the proper role for DOE. Some favor a continuing or larger role, while others favor shifting weapons, waste cleanup, and basic science programs to more logical locations and abolishing the department. Work done to define an appropriate federal role in energy and to justify the continuing bureaucracy and spending has failed to produce satisfying answers. Instead, the objectives that have been identified for DOE tend to be those that in a market economy are achieved more efficiently by the private sector in response to market forces.

A large central government function is particularly difficult to justify in an era when energy supplies are plentiful, prices are moderate, competition is strong, and the key decisions affecting energy are made quite competently by millions of individuals and organizations outside Washington.

Reassessing Vulnerability

Federal officials have justified much government involvement in energy policy and many of DOE's energy programs as being necessary to reduce the nation's "vulnerability" to prolonged oil supply interruptions or sustained high prices. This rationale has persisted and is often repeated by organizations seeking federal subsidies, even though U.S. and world energy markets and geopolitical considerations have changed considerably in recent years. Continuing use of this rationale by the government should, at a minimum, be accompanied by a detailed definition of *vulnerability*, specific criteria for assessing changes in vulnerability, and periodic public disclosure of the basis for the government's conclusions. Objective assessment of the rationale and justification by people with no financial stake in the outcome would lend credibility to the conclusions.

The Economic Impact of the Dollar Outflow for Imported Oil

Without a doubt, the United States experiences a significant outflow of dollars for imported oil, and the quantities of oil imported are likely to increase in years to come. This fact is often cited by government officials and by organizations seeking subsidies or special treatment for development of domestic energy sources. A seldom-mentioned fact, however, is that a significant portion of these outflow dollars for oil flow back to the United States, directly or indirectly, to pay for exports of U.S. goods and services. DOE focuses only on the half of the picture that justifies its policies and programs. It is time to present both sides through a thorough, balanced assessment of the macro- and microeconomic effects of both the dollar outflow and the "recycling" of those dollars.

Conclusions

Government policies and programs adopted during the past twenty years in response to the oil shock and succeeding events provide a rich menu for study by policy analysts. Lessons that could be learned would almost certainly provide insights applicable to many current and proposed policies and programs, not just those involving energy. Two cautions are in order. First, there are no guarantees that even the best analysis will have an effect on policies, because of vested interests and powerful constituencies. My suggestions may reflect a naive view that analysis makes a difference. Second, there may be little funding available. The logical source of support, DOE, may have little interest in analysis that evaluates past policy failures objectively or is critical of current DOE roles and policies. Further, current low levels of public interest in energy probably mean that funding is unavailable from other government agencies or foundations that might otherwise have an interest in objective public policy analysis.

NOTES

1. Energy modelers were neither all-knowing nor infallible. For example, problems with price controls are generally recognized by those completing Economics 101, not just by modelers. Also, modelers' underestimation of the elasticity of energy demand and supply contributed to their incorrect supply and demand forecasts.

2. One occasion in 1971, thirteen in 1973, ten in 1974, five in 1975, and two in 1976.

3. Energy functions, in total, are relatively small compared with DOE's spending for nuclear weapons research, development, testing, engineering, waste cleanup, and basic science support.

10. Twenty Years of Energy Policy

What Should We Have Learned?

DAVID L. GREENE

More than twenty years ago, the Arab members of OPEC launched a boy-cott of oil shipments to the United States in protest of its support for Israel in the October 1973 Yom Kippur War. Very likely, no one was more sur-prised at the result than the Arab OPEC states themselves (Yergin, 1991). Oil prices, which had been declining slowly for decades, doubled within a year, sending the Western economies reeling under inflation, unemployment, and recession. A second oil price shock in 1979–80, associated with the Iran-Iraq War, had a similarly devastating impact on oil-importing countries. But oil prices fell from 1982 through 1985 and collapsed in 1986, falling to levels very near those prevailing before the 1973 oil embargo.

The past twenty years have taught us important lessons regarding the effects of energy price deregulation, the use of technology in response to energy price changes, and the role of strategic petroleum stocks. However, there are two lessons that we may have seriously misunderstood. The first is that the oil price shocks posed very big, very real problems for oil-importing countries. These problems still reverberate throughout the U.S. economy, and OPEC's market power remains a serious threat. Moreover, our failure to understand the impact of cartel behavior has cost us dearly these last twenty years. A recent study estimates the cost of the OPEC price shocks at $4 trillion (Greene and Leiby, 1993).

The second lesson is that energy markets alone cannot solve societal problems resulting from oil dependence. There is continuing need for ef-fective governmental energy policy. I illustrate this larger lesson with the example of automobile fuel-economy standards. Automobile fuel-economy regulation has worked effectively to reduce oil consumption and its exter-nalities. It can also work effectively in the future. Regulation is needed because there are significant externalities resulting from oil use, and the market for energy efficiency will not necessarily produce the most cost-effective level of energy efficiency based on private costs.

To some these lessons may appear self-evident. But to others they may sound like anathema. For example, Bohi and others argue that OPEC did not cause the suffering that followed the oil market disruptions of the 1970s. OPEC is a convenient scapegoat for the failure of excessive government regulation of energy markets (and, indeed, of the entire economy). According to this view, OPEC never was an effective cartel; it merely benefited serendipitously from the disruption of world energy markets caused by misguided policies and the natural order (Bohi 1989; Bohi and Toman, 1992). As for federal fuel-economy standards, they are viewed as a perfect example of exactly the wrong kind of energy policy. Any fuel-economy improvements worth making were made in response to rising gasoline prices. Any that were caused by regulation must have been economically inefficient and therefore unnecessary burdens on automobile manufacturers and consumers (Crandall et al., 1986; Leone and Parkinson, 1990).

These interpretations of events are convenient and comforting. They imply that the energy problem was largely of our own making, and if we can just keep government out of energy markets, we need not fear future energy problems. But these are the wrong lessons to have learned. If we heed them, we will most likely be condemned to repeat the mistakes of history.

Oil Price Shocks: How Big a Problem?

The costs of oil price disruptions have been huge. Greene and Leiby (1993) retrospectively estimated that the economic costs to the United States of OPEC's manipulation of world oil prices amounted to $4 trillion between 1972 and 1991.[1] Sums this large are difficult to comprehend. The costs to the United States of OPEC's oil market monopolization are on the order of one year's GNP ($5.5 trillion in 1990, e.g.), or total expenditures on national defense from 1972 to 1991 ($5.2 trillion). Oil market disruptions are one of our largest national problems of the past twenty years.

Estimates such as this depend on the definition of a counterfactual case: costs compared with what? Our estimated costs were compared with a status quo ante world oil market, one in which neither OPEC nor anyone else exercised monopoly power.[2] During the pre-1973 oil embargo era, oil prices were stable or had gradually declined for a quarter of a century

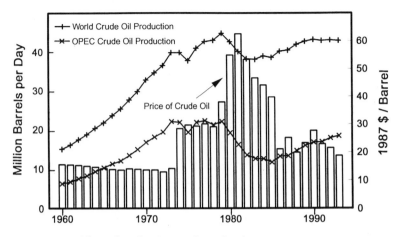

Figure 10.1   **World Crude Oil Prices and Production, 1960–1993**
Sources: U.S. DOE/EIA,1992b: table 68; U.S. DOE/EIA, 1994: tables 9.1 and 10.1b.

(fig. 10.1). What happened in October 1973 was not that oil wells abruptly dried up or that oil discovery and development became unexpectedly more difficult; instead, a group of producers collectively holding three-fourths of the world's oil reserves and accounting for more than half of the world's production suddenly decided to restrict supply in a very tight market. In other words, they exercised monopoly power.

Sudden oil price shocks caused by monopolistic behavior create three types of economic costs for oil consumers (Greene and Leiby, 1993). First, a *loss of potential GNP* occurs because from the economy's perspective, there is no difference between a price increase due to physical scarcity and one due to monopoly behavior. A price increase is a signal that resources have become economically more scarce. Because petroleum products are so well suited for certain uses such as transportation fuels, they have few close substitutes. In a world where oil is more scarce, less output can be produced with the same amount of capital and labor. This loss of the potential to produce is permanent, so long as prices remain high. Conversely, when prices suddenly fall, there is a gain in potential GNP.

Second, *macroeconomic adjustment costs* arise because the economy cannot instantly adjust prices and wages. More important, it cannot adjust the technology embodied in capital to the new price regime. When oil prices rose, energy-using capital became inconsistent with the new price regime: it was too inefficient. If we could have instantly adjusted the efficiency of energy-using capital, we could have driven more, kept our homes warmer in winter, produced more electricity, and manufactured more output, despite higher energy prices. Because it takes time to adjust, the economy suffers an additional loss of GNP above and beyond the "permanent" loss of potential output. As adjustment occurs, this loss disappears over time. Unlike the potential GNP effect, macroeconomic adjustment losses occur even when prices fall.

Third, monopolistic pricing causes a *transfer of wealth* from oil consumers to oil producers. From the perspective of the world economy, this transfer is not an economic loss; it is simply a change in the terms of trade between producers and consumers in favor of the former. World economic output remains the same, but from the viewpoint of the United States as an oil importer, there is a real loss of wealth. Simply stated, we get poorer and they get richer. We assume that this mattered to the United States as a nation, and so we count it as a cost. Wealth transfer is the easiest component to estimate, since it is equal to U.S. oil imports times the difference between the actual market price and the assumed counterfactual (status quo ante) price. Transfer of wealth is the only one of the three components that depends directly on how much oil the United States imports.

At least fourteen estimates of the effects of energy price shocks on U.S. GNP have reached a consensus that a sudden 50 percent increase in world oil prices will produce something like a 2.5 percent drop in GNP (Hickman, 1987). That is, the elasticity of GNP with respect to oil price is about -0.05. This includes *both* the "permanent" loss of potential GNP and macroeconomic adjustment costs. Although there are few estimates of the individual components, it has been suggested that they are roughly equal in magnitude (Pindyck, 1980). The estimates closely agree with economic theory.[3]

Using historical data, the GNP elasticity of -0.05, and appropriate methods for reflecting dynamic adjustments over time, Greene and Leiby (1993) estimated that the total costs of oil price shocks to the U.S. economy from 1972 to 1991 amounted to more than $4 trillion in undiscounted 1990 dollars (fig. 10.2).[4] Their estimates break down as follows:

—loss of potential GNP: $2.1 trillion ($3 trillion present value at 5 percent real discount rate)
—macroeconomic adjustment losses: $800 billion ($1.4 trillion present value)
—wealth transfer: $1.2 trillion ($1.9 trillion present value)

No one argues that these costs are trivial. An important issue, however, is whether such costs can be avoided (Bohi and Toman, 1992). Equally impor-

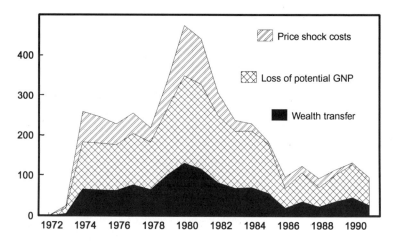

Figure 10.2   The Costs of Oil Market Monopolization to the U.S. Economy, 1972–1991 (billions of 1990 dollars)
Source: Greene and Leiby, 1993.
Note: Macroeconomic adjustment cost share, 50 percent; rate parameter, 0.85.

tant is whether such oil price shocks are likely to happen again. To understand these issues, we need a theory of how the world oil market works.

Oil Market Monopoly: Can It Happen Again?

Few doubt that OPEC has exercised monopoly power in the world oil market. Even fewer doubt that OPEC nations could again exercise monopoly power if they were willing and able to cooperate (Bohi and Toman, 1992). But many are confused about the nature of OPEC's market power and cannot understand how a cartel that can send oil prices skyrocketing one year can lose control of the market only a few years later. To some this is evidence of a failure to use market power, or even of a total collapse of the cartel. In fact, it reflects the economic constraints on the cartel's monopoly power. I shall now outline a theory explaining how OPEC uses its market power, what the limits of that power are, and how the actions of the United States and other oil-consuming nations affect it.

I begin with the counterintuitive assumption that oil is not a depletable resource. No one has explained this paradox better than Adelman (1989: 19) in his description of the history of U.S. oil reserves following World War II: "No area in the world is as drilled-up today as this country was (excluding Alaska) in 1945; 'Remaining recoverable reserves' were 20 billion barrels. In the next 42 years, the 'lower 48' produced not 20 but 100 billion, and had 20 billion left. Equally important, there was no increase in real cost before 1973." Were these stable costs and reserves of 100-plus billion barrels a miracle, like Moses striking the desert rock to get water? Hardly. The lesson is that oil reserves are not a fixed stock to be allocated over time, but an inventory, constantly consumed and replenished by investment.

Oil resources are ultimately finite, but so are resources of iron and aluminum. The point is that we are so far from the point of exhausting oil that the market behaves as if oil resources were infinite. As Mabro (1992: 3) put it, "The geophysical limits may bite one day, but this day of reckoning is so far ahead as to have, on any conceivable assumption about discount rates, no impact on price."

Thus, a static model of oil supplier behavior explains the nature of monopoly power in the world market. The basic theory of monopoly pricing predicts that a complete monopolist will charge a premium over the cost of production that depends on the elasticity of demand (β). The ratio of the monopolist's revenue-maximizing price (P) to the cost of production (C) is as follows:

$$P/C = 1/[1 + (1/\beta)] \tag{10.1}$$

OPEC, of course, is not a complete monopolist. Although OPEC countries contain three-fourths of the world's reserves, in 1990 they produced only 43 percent of world supply (U.S. DOE/EIA, 1994: table 10.1b). According to economic theory, a monopolist that controls only part of the market will also charge a markup over the cost of production. This partial-monopoly markup is dependent on the supply responsiveness of other competitive producers and on the partial monopolist's market share. If we define the supply response (δ) to be the number of barrels produced by the rest of the world in response to a one-barrel-per-day change in OPEC production, and let σ represent OPEC's market share, then OPEC's monopoly price markup is described by the following equation:

$$P/C = 1/[1 + (\sigma/\beta)(\delta + 1)] \tag{10.2}$$

The monopoly price markup specifies the price that maximizes the monopolist's total profit. If the monopolist charges more than P, demand will drop and total profits will fall. If it charges less, profits will also fall. Equation (10.2) says that a partial monopolist's market power will increase if demand elasticity decreases, supply response decreases, or market share increases.[5] In interpreting equation (10.2), we note that β is less than zero,

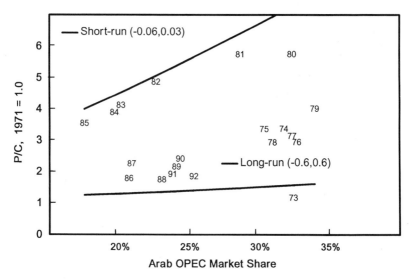

Figure 10.3   Oil Prices and Arab OPEC's Market Share
Source: Greene, 1991.

σ is greater than zero and less than 1, and, usually, δ is less than zero and greater than −1. As market share (σ) goes to zero, as supply response (δ) approaches −1 (if OPEC reduces production by one barrel, the rest of the world replaces that with a full barrel), and as demand (β) becomes infinitely elastic, OPEC's market power disappears (*P/C* approaches 1).

A fact of absolutely fundamental importance to world energy markets is that short-run responses are far more restricted than long-run responses. Short-run elasticities (those pertaining to periods of one year or less) of demand and supply for oil are on the order of −0.06 and 0.03, respectively. Long-run elasticities are at least ten times as great, for example, −0.6 and 0.6 (Greene, 1991).[6] In light of equation (10.2), this means that OPEC can charge a price in the short run many times higher than it can sustain in the long run. In the long run, demand responsiveness will increase, the rest of the world's supply response will increase, and OPEC will lose market share. As market share erodes, so does short-run market power. This simple model provides powerful insight into oil market events of the past twenty years.

Consider that for given values of demand and supply elasticities, *P/C* or market power is a function of market share, σ. If we graph *P/C* as a function of market share using first short- and then long-run elasticity values, we get two curves representing the profit-maximizing monopoly prices that can be achieved in the short run and those that can be sustained in the long run (fig. 10.3).[7] There is a wide gap between the two that expands at an increasing rate as market share increases. Any point between the curves is in some sense profit-maximizing, depending on whether OPEC takes a long- or short-run perspective.

The precise location of these profit-maximizing curves is unknown. When one adds to this the difficulty of reaching a consensus about how much profit to make in the short run versus the long run, among other issues, it is likely that the cartel will opportunistically explore the space between the short- and long-run curves, learning the market as it goes. Figure 10.3 shows short- and long-run curves for the Arab members of OPEC rather than all of OPEC. This representation is more consistent with the view that it is a core of OPEC members rather than all of OPEC that functions as a monopolistic cartel.

In figure 10.3, historical prices, identified by the last two digits of the year, are plotted versus the actual Arab OPEC market share in that year. These points form an interesting pattern. In 1973, world oil prices appear to be below the long-run monopoly price curve. In 1974, prices jump above the curve, and they remain in virtually the same place for five years. In a growing market, some price above the long-run static equilibrium price should be sustainable at constant market share. In 1979 and 1980, prices again leap upward. But this time Arab OPEC members immediately begin losing market share. As market share declines for the next six years, price continues to be held at a level near the long-run monopoly curve, ensuring continued loss of market share. Finally, in 1986, no longer able to divide a shrinking pie satisfactorily, the cartel abandons its defense of price, allowing it to fall to near the long-run monopoly price curve. At this level, growing oil demand from a growing world economy allows OPEC to regain lost ground.

Is OPEC dead? The analysis of OPEC as a partial monopoly suggests that it is only sleeping until the growth of oil demand restores its market share and power. This is virtually certain to happen, since OPEC members hold three-fourths of the world's known oil reserves. The history illustrated in figure 10.4 suggests the possibility but not necessity of endless cycles of price shocks and collapses. Such a pattern may even provide the greatest revenue to oil producers. Figure 10.4 also shows the U.S. Department of Energy's projection of OPEC's market share for 2010. In light of history, we would be fortunate indeed if OPEC agreed to produce so much oil without taking advantage of the opportunity to charge much higher prices.[8] By the year

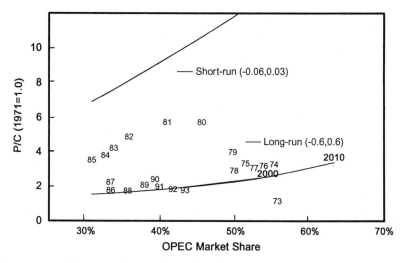

Figure 10.4   **Oil Prices and OPEC's Market Share**
Sources: U.S. DOE/EIA, 1992a; 1992b; 1993; 1994.

2010, it appears that OPEC will have both the opportunity and the motive to at least double or triple oil prices. The question we must answer is, Why wouldn't they do so?

The implications of this analysis for energy policy are clear, but not easy to carry out. Oil-consuming nations should strive to increase the short-run elasticities of demand, increase their ability to produce oil and oil substitutes, and restrain OPEC's market share. Increasing demand elasticity means developing substitutes for oil and oil-based activity. To date we have been relatively unsuccessful at this. Although the building and industrial sectors of the economy have reduced their dependence on petroleum since 1973, oil as a share of industrial and transportation energy use has remained nearly constant. Because these sectors dominate petroleum use, oil's share of total U.S. energy use has decreased only 7 points (from 47 percent to 40 percent).

It is also reasonable to expect that price decontrols on all forms of energy will increase short-run price elasticity. In addition, we have amassed the Strategic Petroleum Reserve to provide a short-run surge of oil supply replacement, but we have not yet shown that we know how to use it. There are some positive developments to point to, but nothing that seems large enough to seriously threaten OPEC's market power once its market share climbs above 50 percent.[9]

What about alternative fuels? For the past seven years the U.S. Department of Energy (DOE) has carefully investigated the technology, vehicle and infrastructure costs, resources, and production costs for alternative transportation fuels. DOE has found that, currently, the closest substitutes for conventional petroleum fuels are compressed or liquefied natural gas and methanol derived from natural gas.[10] There is a great deal of gas worldwide, and the ability to use it as a feedstock for transportation fuels could, in principle, help weaken OPEC's market power. The gas most likely to be used to produce fuels for export is gas unassociated with petroleum deposits, and for which there is no substantial regional market. DOE has examined world resources of unassociated, undeveloped natural gas and concluded that there is a great deal available that *could* be produced at prices competitive with petroleum (U.S. DOE/ODIEP, 1993). The bad news is that these gas resources, like oil resources, are highly concentrated in a few countries. Even worse, with the exception of the countries of the former Soviet Union, the countries with the largest unassociated gas resources are also OPEC members (fig. 10.5). This does not guarantee that methane-based fuels will not help energy security, but it does shift the burden of proof, and the results are not yet in.

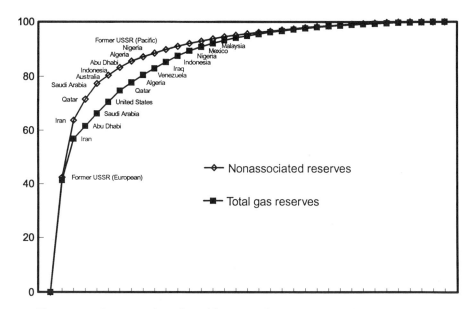

Figure 10.5   Concentration of World Reserves by Country, 1991 (percent)
Source: U.S. DOE/ODIEP, 1993: table S-1.
Note: Countries ranked by size of reserves.

Energy-Efficiency Regulation: Why Does It Work?

One thing we can do to weaken OPEC market power and improve energy security is to reduce demand for petroleum by improving energy efficiency via technological change. One successful national energy policy that achieved significant and cost-effective reductions in U.S. petroleum demand was the federal automotive fuel-economy standards (AFES), better known as the corporate average fuel-economy, or CAFE, standards. CAFE standards required all major manufacturers to increase their sales-weighted average fuel economy from levels in the vicinity of 15 miles per gallon (MPG) in 1975 to 18 MPG by 1978, with annual increases thereafter until the target of 27.5 MPG was reached in 1985. Standards post-1985 were to be 27.5 MPG unless raised or lowered by federal rule making. Light-truck standards were also established, but the Department of Transportation was given considerable latitude to set the timetable and levels. Perhaps as a result, the light-truck standards were not as ambitious as passenger-car CAFE standards. A substantial fine was imposed for failing to meet the standards: $5 per car sold per 0.1 MPG of shortfall. Manufacturers were allowed to earn credits by exceeding the standards and could carry this credit forward or backward for three years.

It is indisputable that manufacturers met the standards. There is considerable argument, however, about whether compliance was due to the standards themselves or to market forces driven by higher gasoline prices. There is also dispute over whether, on balance, the standards were cost-effective. That standards rather than fuel prices constrained manufacturer behavior seems both self-evident and overwhelming. Even a cursory look at the MPG trend over time (fig. 10.6) would seem to shift the burden of proof to those arguing for market forces. This issue has been examined in depth elsewhere (Greene, 1990).

The issue of cost-effectiveness is more difficult to resolve, since the impact of changes in vehicle designs on consumer satisfaction must be accounted for as well as financial costs, and since the counterfactual case is difficult to define. Again, casual observation suggests that the cars of today are far better than the cars of twenty years ago, and more rigorous analysis supports this impression (Greene and Liu, 1988). In any case, it is not my intention to prove that CAFE has been effective; my point is, rather, to explain why energy-efficiency regulations can and do work.

Fuel-economy regulations are well suited to the way the market system determines automotive fuel economy. The costs of oil use not reflected in

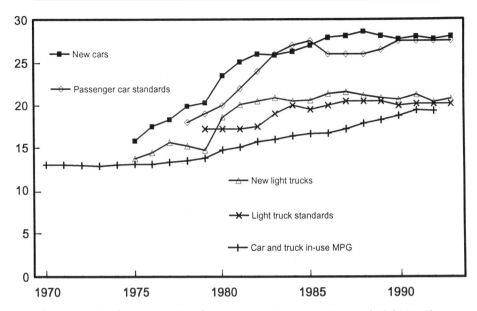

Figure 10.6   **Fuel-Economy Trends, 1970–1993, Passenger Cars and Light Trucks (EPA combined MPG)**
Sources: New-vehicle MPG–Murrell, Hellman, and Heavenrich, 1993: table 1; in-use vehicle MPG–U.S. DOT/FHWA, 1993: table VM-1.

market prices (e.g., energy security, environmental damages) justify some intervention by government in the oil market. Economic theory shows that in a competitive market where consumers and producers make utility- and profit-maximizing decisions, a tax equal to the marginal external cost of consuming a barrel of oil will generally be more efficient than regulating oil use or energy efficiency (Baumol and Oates, 1989). Unfortunately, the market for fuel-economy bears little resemblance to the perfect, competitive market of theory. We misunderstand the impacts of fuel-economy regulation not because of "market failure" but because of "theory failure." The simple competitive-market model fails because it is not a reasonable representation of consumer decision making about automotive fuel-economy (the algorithm is wrong), and it does not take into account the important time lags, dynamics, and risks involved in manufacturer decision making about fuel-economy.

The nature of the trade-off between fuel-economy improvement and initial vehicle cost is such that, over a wide range of values of MPG, the *net* benefit to the consumer is rather small. This is the most critical point. Improving vehicle fuel economy while maintaining other vehicle characteristics of value to the consumer (e.g., performance, capacity, comfort, amenities) generally requires the use of technology that costs a little more. A multipoint fuel-injection system costs more than a carburetor, a five-speed automatic transmission with torque converter lock-up costs more than a three-speed without, four valves per cylinder cost more than two, radial tires cost more than bias ply, and so on. In return for this greater capital expenditure, the motorist gets a time stream of savings in fuel costs. The *net* benefit to the consumer is the difference between the two. As long as the initial technology cost is about the same size as the discounted present value of fuel savings to the consumer, the net benefit will be small. Small net benefits to even fairly large fuel-economy improvements generally accrue for automotive technology, and probably for other technology as well. This should not be surprising. It depends only on the existence of a substantial amount of energy-efficiency technology reasonably close to being cost-effective but that does not quite pass the market test.

There is substantial evidence for the existence of nearly cost-effective fuel-economy technology. A report by the National Research Council (NRC, 1992) contains curves describing the total cost of technology-based fuel-economy improvement for four market segments each of passenger cars and light trucks. Figure 10.7 shows the cost of fuel-economy technology for a typical compact car as well as the present value of fuel savings.[11]

The graph assumes a gasoline price of $1.42 per gallon, a projected price for the year 2005 in which the fuel-economy improvements were assumed to be implemented. Savings are discounted at a real rate of 10 percent per year over twelve years. From a base of 28.5 MPG, the cost of a 2-MPG increase is about $100, a 4-MPG increase costs about $250, and an 8-MPG increase costs about $600. At a fuel price of $1.42 per gallon, the present value of benefits exceeds costs up to about 36 MPG, at which point the net private value becomes negative. In the competitive-market model, 36 MPG would not be the market solution, since it does not maximize net benefit to the consumer. The optimal market solution is in the vicinity of 31 MPG, a good 5 MPG lower.

The important point is that the net value to the consumer is plus or minus $100 over the range from 28.5 to 36.5 MPG. Consider the significance of $100 in the automobile purchase decision. It is substantially less than the cost of most options, on a par with the cost of wheel covers or floor mats. With an average car costing about $17,000, the net value of fuel economy is on the order of half a percent of the purchase price. Because fuel economy is but one of many factors the consumer considers (e.g., price, size, safety,

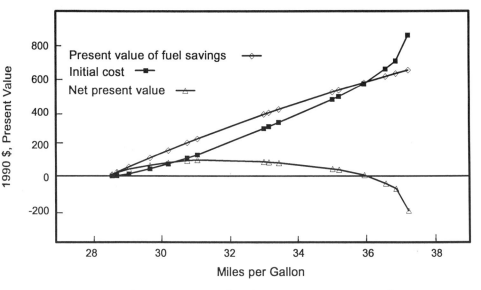

Figure 10.7   Net Value to Consumer of Fuel-Economy Improvements for a 28.5-MPG Compact Car, Gasoline at $1.42 per Gallon
Source: NRC, 1992: compact cars curve "B."

performance, reliability, styling, color, stereo system, transmission, engine, tires, lights, mirrors), it is highly unlikely that the consumer will actually *optimize* his fuel-economy decision. Consider that you have never calculated the present value of fuel savings for automobiles you were considering buying, and you don't know anyone who has. Even if you had, what fuel prices would you have used for the future? Would you believe the fuel-economy numbers on the car sticker? (DOE surveys indicate that most car buyers do not read the MPG sticker or the fuel-economy pamphlets.) Among the dozens of characteristics car buyers consider, fuel-economy is highly unlikely to be determined by optimization. It is determined by some other algorithm, like "satisficing," or it is simply ignored. As figure 10.7 shows, for all practical purposes the average car buyer will be indifferent about fuel economy over the entire range of MPG.

From the manufacturers' perspective, to get to 36 MPG, essentially the full list of fuel-economy technologies must be used, resulting in a completely new vehicle redesigned from the ground up (e.g., new engines, new transmissions, new body design for aerodynamics, new tires, new accessories). And this means not just one new vehicle. If fleet fuel economy is to improve significantly, *all* product lines must be completely redesigned. It means new engine lines, each with a normal life of fifteen years or so (NRC, 1992). It means new component and body manufacturing equipment, capital usually turned over every ten years or so. It means enormous investments in engineering and capital over a five- to ten-year period—all for a change about which customers are indifferent.

And what if customers don't like the changes for some reason? What if fuel prices don't rise to $1.42 per gallon? And what if your competitors don't follow your lead? Of course, the optimal market solution (31 MPG) requires only part of these massive changes, but it is still likely to get a big yawn from car buyers. Better to focus limited resources on what paint colors are in, or better advertising, or improved quality control, or airbags. From the manufacturers' perspective, meaningful fuel-economy improvement entails significant risk for insignificant reward.

From the societal viewpoint, the issue looks entirely different. Not only would it be nice to get to 31 MPG, presumably the economically efficient point, but externalities like energy security, greenhouse gas emissions, and hydrocarbon emissions (DeLucchi, Greene, and Wang, 1994) are likely to justify going all the way to 36 or 37 MPG (e.g., Greene and Duleep, 1993). From the societal viewpoint, *if* standards are set at a reasonable value (say 36 MPG) and *if* sufficient time is allowed for manufacturers to plan and adapt (say ten to fifteen years), then significant social costs can be avoided at very little cost.

From 1975, the year in which the Energy Policy and Conservation Act of 1975 established fuel-economy standards for passenger cars and light trucks, to 1985 the fuel economy of new cars increased by over 75 percent, and the fuel economy of light trucks improved by more than 50 percent (fig. 10.6). As a result, fleet fuel economy improved from 13 MPG in 1975 to 19 MPG in 1991. But light-duty-vehicle fuel economy has not increased since 1985. In 1992, for the first time in two decades, in-use fuel economy of cars and light trucks declined.

Conclusions

Twenty years after the oil price shock of 1973–74, it is all too easy to convince ourselves that things are different. Oil prices are low, and oil is plentiful. OPEC is in disarray. We have the Strategic Petroleum Reserve. We have deregulated gas markets and have abolished price controls on petroleum. A petroleum futures market has been established. But have these actions altered the fundamental nature of world energy markets? Has the distribution of resources changed? Have the short- and long-run elasticities of demand and supply changed?

OPEC has allowed prices to crash, and its members are fighting among themselves. Surely they will never again intimidate oil-consuming nations. But what will happen when rising world oil demand in developing and developed countries turns increasingly to OPEC for oil and OPEC's share of the world market passes 50 percent on its way upward? There is little doubt that world energy markets are headed in that direction, and that in the first decade of the twenty-first century, world demands on OPEC production will strain the cartel's capacity to produce (Hogan, 1992; *Economist*, 1994). Has the distribution of oil reserves changed fundamentally, or doesn't OPEC still hold three-fourths of the world's known oil reserves? When the opportunity arises, will OPEC countries choose stability in world oil markets, or will they take the opportunity, however short-lived, to acquire fantastic wealth?

And what about natural gas? Will gas and the alternative fuels that can be made from it, such as methanol, change the fundamental equation? Although we have yet to see a large-scale world market in fungible natural gas, it appears that the world's exportable gas resources are as regionally concentrated as oil resources, and that OPEC countries and the former Soviet Union dominate. It remains to be seen whether oil and gas producers would allow gas sales to undermine oil prices.

We have doubled the efficiency of new passenger cars, increased light-truck MPG by two-thirds, and achieved similar gains in air and rail transport. However, new-car efficiency has not increased appreciably in the past ten years, and in the rest of the transportation sector, energy-efficiency improvement is slowing to a halt. And despite the promise of alternative fuels, the transportation sector remains nearly totally dependent on petroleum. In 1973, petroleum accounted for 96 percent of the energy used to power transportation; in 1993, it accounted for 97 percent of transportation energy (fig. 10.8). In 1973, total U.S. oil expenditures amounted to 2 percent of GNP; in 1991, the proportion was unchanged (U.S. DOE/EIA, 1992b: tables 52, 71, C1). In 1973, the United States imported 35 percent of its oil; in 1993, the figure was 44 percent, the highest import level on record save for 1977 (47 percent) (U.S. DOE/EIA, 1994: table 1.8).

Things have changed, are changing, and will continue to change. But we cannot ignore the risk of future oil price shocks. There are no grounds for complacency. Oil price shocks cost the United States dearly in the past and could cost us dearly in the future. Despite twenty years of experience, we are

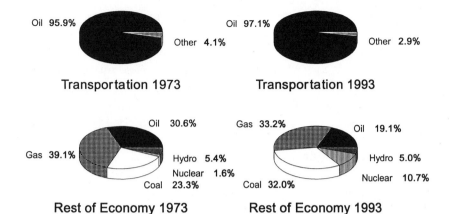

Figure 10.8 The Continuing Oil Dependency of the Transportation Sector, 1973 and 1993

Source: U.S. DOE/EIA, 1994: tables 2.3 and 1.4.

only a little better prepared now than we were twenty years ago. More discouraging yet, we seem to want to pretend that it can't happen again and that it really didn't happen at all, at least not the way it seemed at the time. If those who fail to learn from history are condemned to repeat it, we must be banking on being graded on a curve.

NOTES

The author is a senior research staff member at the Center for Transportation Analysis, Oak Ridge National Laboratory, managed by Lockheed Martin Energy Research, Inc., for the U.S. Department of Energy. This chapter has been authored by a contractor of the U.S. government under contract no. DE-AC05-84OR21400. Accordingly, the U.S. government retains a nonexclusive, royalty-free license to publish or reproduce the published form of this contribution, or allow others to do so, for U.S. government purposes.

1. The $4 trillion estimate is a simple (undiscounted) sum of constant 1990 dollars. Costs in each year were inflated to constant 1990 dollars using the GNP implicit price deflator but were not converted to present value.

2. One could argue that the "Seven Sisters" (the world's dominant oil companies of the pre-OPEC era) exercised some form of monopoly in world oil markets. If one believes this argument, then our counterfactual case should be interpreted as status quo ante rather than unmonopolized.

3. Bohi (1989) derived a theoretical upper bound on the impact of oil price on GNP that he believed showed that the loss of potential GNP could not possibly be as large as the macroeconomic models indicated. His conclusion, however, was due to a calculation error (see Greene and Leiby, 1993: 22, n. 15). When computed correctly, Bohi's calcula-

tion indicates an upper bound on the potential loss of GNP that is generally larger than the impacts predicted by the macroeconomic models.

4. If one assumes that oil prices would otherwise have increased at a real rate of 2 percent per year, the total loss decreases only $0.5 trillion, to $3.6 trillion.

5. If market power is the ability to charge a higher-than-competitive price, then market power is aptly measured by the monopoly price markup.

6. A more detailed mathematical exposition can be found in Greene (1991), along with an explanation of the relationship between competitive supply response and the elasticity of supply.

7. In fact, the short-run curve is undefined unless one allows price elasticities to increase as price increases, as Greene (1991) has shown. Increasing price elasticity with increasing price is certainly accurate, since drastic oil price increases eventually reduce demand by throwing economies into recessions.

8. For a discussion of the implications of such high levels of production for the necessary expansion of OPEC production capacity, see Hogan (1992).

9. OPEC market shares of 50–60 percent imply core market shares of over 30 percent.

10. Propane is also a reasonable substitute fuel, but it is derived from crude oil or natural gas production.

11. The technologies used to achieve these improvements are described in Appendix B of the National Research Council (1992) report on automotive fuel economy, and the method used to create the fuel-economy supply curve is described in Appendix E.

REFERENCES

Adelman, Morris A. 1989. "Mideast Governments and the Oil Price Prospect." *Energy Journal* 10 (2): 15–24.

Baumol, William J., and Wallace E. Oates. 1989. *The Theory of Environmental Policy.* 2d ed. Cambridge: Cambridge University Press.

Bohi, Douglas R. 1989. *Energy Price Shocks and Macroeconomic Performance.* Washington, D.C.: Resources for the Future.

Bohi, Douglas R., and Michael A. Toman. 1992. *Energy Security Externalities and Policies.* ENR 92-05. Washington, D.C.: Resources for the Future, May.

Crandall, Robert W., et al. 1986. Chap. 6 of *Regulating the Automobile.* Washington, D.C.: Brookings Institution.

DeLucchi, M. A., David L. Greene, and M. Q. Wang. 1994. "Motor-Vehicle Fuel Economy: The Forgotten Hydrocarbon Control Strategy?" *Transportation Research A,* 28A (3): 223–44.

The Economist. 1994. "Power to the People: A Survey of Energy," June 18, pp. 3–18.

Greene, David L. 1990. "CAFE or Price? An Analysis of the Effects of Federal Fuel Economy Regulations and Gasoline Price on New Car MPG, 1978–1989." *Energy Journal* 11 (3): 37–57.

———. 1991. "A Note on OPEC Market Power and Oil Prices." *Energy Economics* 13 (2): 123–29.

Greene, David L., and K. G. Duleep. 1993. "Costs and Benefits of Automotive Fuel Economy Improvement: A Partial Analysis." *Transportation Research A,* 27A (3): 217–35.

Greene, David L., and Paul N. Leiby. 1993. *The Social Costs to the U.S. of Monopolization of the World Oil Market, 1972–1991.* ORNL-6744. Oak Ridge, Tenn.: Oak Ridge National Laboratory, March.

Greene, David L., and J. T. Liu. 1988. "Automotive Fuel Economy Improvements and Consumers' Surplus." *Transportation Research A,* 27A (3): 203–18.

Hickman, Bert G. 1987. "Macroeconomic Impacts of Energy Shocks and Policy Responses: A Structural Comparison of Fourteen Models." In *Macroeconomic Impacts of Energy Shocks,* edited by Bert G. Hickman, Hillard G. Huntington, and James L. Sweeney. New York: Elsevier Science Publishers, North-Holland.

Hogan, William W. 1992. "Oil Market Adjustments and the New World Order." Discussion paper. Harvard Center for Business and Government, John F. Kennedy School of Government, Harvard University, Cambridge, March.

Leone, R. A., and T. Parkinson. 1990. *Conserving Energy: Is There a Better Way? A Study of Corporate Average Fuel Economy Regulation.* Cambridge, Mass.: Putnam, Hayes, and Bartlett, May.

Mabro, R. 1992. "OPEC and the Price of Oil." *Energy Journal* 13 (2): 1–17.

Murrell, J. D.; K. H. Hellman; and R. M. Heavenrich. 1993. *Light-Duty Automotive Technology and Fuel Economy Trends through 1993.* EPA/AA/TDG/93-01. Ann Arbor, Mich.: U.S. Environmental Protection Agency.

National Research Council, Energy Engineering Board, Committee on Automobile and Light Truck Fuel Economy (NRC). 1992. *Automotive Fuel Economy: How Far Should We Go?* Washington, D.C.: National Academy Press.

Pindyck, Robert S. 1980. "Energy Price Increases and Macroeconomic Policy." *Energy Journal* 1 (4): 1–20.

U.S. Department of Energy, Energy Information Administration (U.S. DOE/EIA). 1991. *International Energy Annual 1991.* Washington, D.C.

———. 1992a. *Annual Energy Outlook.* Washington, D.C.

———. 1992b. *Annual Energy Review 1991.* DOE/EIA-0384(91). Washington, D.C., June.

———. 1993. *Annual Energy Review 1992.* Washington, D.C.

———. 1994. *Monthly Energy Review.* DOE/EIA-0035(94/05). Washington, D.C., May.

U.S. Department of Energy, Office of Domestic and International Energy Policy (U.S. DOE/ODIEP). 1993. *Technical Report Nine: Development Costs of Undeveloped Nonassociated Gas Reserves in Selected Countries.* DOE/EP-003P. Washington, D.C., January.

U.S. Department of Transportation, Federal Highway Administration (U.S. DOT/FHWA). 1993. *Highway Statistics 1992.* FHWA-PL-93-023 Washington, D.C.: U.S. Government Printing Office.

Yergin, Daniel. 1991. *The Prize: The Epic Quest for Oil, Money, and Power.* New York: Simon and Schuster.

PART THREE

ENERGY AND ENVIRONMENT
The Inevitable Alliance

Society's need to balance energy production and consumption with environmental protection is likely to rage most profoundly in developing nations. Moreover, the process of balancing these two issues will require a role for governments as well as markets. There is an emerging consensus that while the private sector is not always responsive to the need to encompass the negative externalities of energy use, command and control regulation is not always an effective remedy for this shortcoming.

Since 1973, developing countries have become increasingly important consumers of energy—a trend that is likely to continue into the next century, as Dennis Anderson of the World Bank argues in the first chapter in this part. As these countries grapple with many of the environmental challenges stimulated by that energy use, their efforts will continue to prompt analysts to ponder two questions: (1) Are we getting better at managing the environmental impacts of energy production and use through national law and international agreement? And (2) can we develop energy supplies that are both economically and environmentally sustainable, particularly in those developing countries where energy use is growing most rapidly?

Although developing countries are undertaking energy-efficiency efforts, nevertheless their rates of energy consumption continue to increase. As energy consumption increases, so too do rates of pollution. Thus, the paramount need of these countries is energy-efficient, low-polluting technologies.

Developing nations have weak governments unable to undertake comprehensive energy planning while simultaneously protecting precious resources. Exacerbating this problem, many of these countries often claim to be at the mercy of stronger and more highly resource- and technology-endowed nations in international fora that distribute aid, fund development projects, and transfer technologies. Anderson argues that assistance through institutions such as the Global Environment Facility—co-managed by the World Bank, the United Nations Development Program, and the United Nations Environment Program—do not constitute costs but are an investment for the entire planet. To the extent that transfers of technology can help to reduce the volumes of greenhouse gases emitted by developing countries, for example, every nation benefits from abating global warming.

Programs must be developed to produce economical, environmentally benign energy technologies at reasonable cost. These programs must not sacrifice economic growth, however, because lower rates of economic growth do not necessarily lead to lower energy consumption or fewer environmental problems. In fact, they may produce just the opposite. Poorer societies tend to have higher population growth rates and are more dependent on highly polluting biofuels such as dung, crop residues, and fuel wood. The key to reducing environmental disruptions caused by energy production and use in these countries is to create sustainable economic activities, and to promote lower fertility rates and lower infant mortality by improving infant and maternal health care.

Richard Morgenstern, formerly with the U.S. Environmental Protection Agency, the second contributor to this part, is sympathetic to this argument. But he cautions that we must take care to avoid the past mistakes in lending assistance to developing nations. While technology transfer and state-supported technology development are appropriate, governments should not be asked to choose "winners" and "losers." When government makes such decisions, the outcomes are likely to be inefficient and even impractical.

If the goal is to hasten market penetration of small-scale renewable energy technologies that are affordable and feasible, is it not better, Morgenstern asks, to set performance standards for new technologies and let the marketplace design the technologies? The alternative—subsidizing some technologies at the expense of others—may have the inadvertent effect of promoting inefficient, irrelevant, and unpopular technologies that never catch on. One way to hasten such performance standard-setting is to develop a "green GDP" index to highlight those activities that detract from, and those that enhance, environmentally appropriate and desirable activity. Such an index could go far in encouraging—through government-disseminated information—improvements to energy decision making.

Robert Bohm of the University of Tennessee, veteran of another international energy symposium held over a decade ago, concludes this examination of the tensions between environment and energy. Bohm suggests that one area where considerable progress has been made since the 1970s has been in recognizing that the choices before us need no longer be viewed as divergent, radical, or irreconcilable. Neither "hard" nor "soft," centralized nor decentralized, small- nor large-scale energy projects and programs alone will suffice to satisfy growing energy demand while keeping the environmental impacts of energy use in check. Eclectic solutions are required, both because technological solutions have not been as easy to identify as some visionaries thought was possible ten or even twenty years ago, and because we now know that political compromise is endemic to durable decisions.

Bohm reminds us that one reason for optimism today is that unlike ten years ago, when international energy decisions were inextricably tied to Cold War challenges, now at least we are at the point where international energy decisions need no longer be chained to geopolitical rivalries. While the challenges of cooperation are not insurmountable, they will, of course, remain difficult to resolve because developed and developing countries will continue to have divergent notions of equity in energy use and in responsibility for current global environmental problems.

Ironically, the three contributors to this part—while approaching the subject of environmentally sustainable energy development from divergent perspectives—share a common expectation: progress will be made in international negotiations aimed at reducing atmospheric concentrations of greenhouse gases, and developed and developing nations will agree on a process for the equitable transfer of resources to make possible the more efficient, less polluting use of energy in the latter.

In short, Morgenstern (who favors markets over regulation) and Bohm (who is optimistic about prospects for international cooperation due to the end of the Cold War) find Anderson's prescriptions viable, if for different reasons. Moreover, all three authors offer insights into the viability of market-based, decentralized, energy-technology-export strategies of value to policymakers, especially in light of initiatives to develop and export economical, environmentally benign energy technologies to developing countries such as the 1992 United Nations Framework Convention on Climate Change (FCCC), negotiated at the Earth Summit in Rio de Janeiro, and the Global Environment Facility (GEF). There are three scenarios international negotiations toward reducing greenhouse gases may follow as a result of the FCCC.

The first scenario is a realistic, even pessimistic path predicated on the assumption that developed nations will resist demands to transfer energy technologies or additional resources to developing countries. Developed nations control the world's financial instruments and, as Stephen Krasner points out, will probably continue to control the assets of such organizations as the World Bank—a critical part of the GEF—and various multinational aid agencies (Krasner, 1985). Developing countries, for their part, will use international fora such as the Conference of the Parties to the FCCC, which had its first post-Rio meeting in Berlin in 1995, to air grievances against wealthier, industrialized "Annex 1" signatory countries to the FCCC.

This scenario has been widely depicted in some popular accounts of post-Rio discussions, particularly those that take as their point of departure the wide gap between aspiration and expectation apparent in nation-to-nation talks since 1992. They usually allude to the lack of a hard and fast "pressing deadline" in the FCCC for action to reduce greenhouse gases (Bodansky, 1995: 3; Kinley, 1995).

This scenario is also consistent with so-called realist conceptions of the effectiveness of "fuzzy," highly idealistic international agreements. In the realist view, international compliance is inhibited by the ambivalent role of national self-interest and risk perception in the calculations of signatory countries to international agreements (Krasner, 1985). Developed nations are therefore unlikely to take stringent actions to comply with targets or timetables. In fact, they will resist their very introduction. Developing countries, meanwhile, will use various fora established under the convention (e.g., the Conference of the Parties to the FCCC) to air longstanding grievances against developed nations about the need to reallocate finances, to adopt hard and fast targets and timetables for reduction, and to include developing nations in the scientific assessments of climate change impacts.

Furthermore, according to this scenario, wealthier nations, through their dominance of global financial institutions and their far better ability to adapt to the adverse impacts of global warming, will put the onus of reducing future emissions of greenhouse gases on developing countries seeking aid and most-favored-nation trading status. Post-Rio negotiations aimed at establishing a durable compliance regime for the FCCC, and moving to a more rigorous set of protocols to reduce emissions after the year 2000, provide ample evidence to support this scenario (*United Nations Climate Change Bulletin*, 1995). Developed-country commitments to "new and additional" resources (as required by the FCCC) have been lacking, and in the absence of firm measures to reduce emissions, many Annex 1 nations admit that their emissions of greenhouse gases are likely to *increase* between 1990 and 2000—not *stabilize*, as was part of the original FCCC commitment (Kinley, 1995).

In contrast to this pessimistic scenario, however, is a far more optimistic view. Some observers argue that developed countries must move—and are moving, slowly—toward "leveling the playing field" for developing countries out of regard for the imminent threat of climate change. Structural changes in the international economic and political order can be brought about if the political will to do so is manifest.

To achieve this scenario, however, several things must fall into place. First, according to such figures as Maurice Strong, the director of the United Nations Conference on Environment and Development at the time the FCCC was negotiated, climate change and other global environmental issues must come to be viewed as highly salient by a majority of citizens in a majority of countries, and achieving justice must become the cornerstone of international agreements. In addition, individual countries must be willing to develop honest and transparent plans to demonstrate compliance with agreements designed to reduce greenhouse gas emissions. Recent evidence suggests that this goal does appear to be achievable, that national governments can indeed be persuaded to focus upon long-term risks not in their immediate interest—though proponents of this goal would hardly abandon realist expectations regarding the importance of monitoring compliance and deterring "free riders" (Benedick, 1991; Mitchell, 1994). Finally, international institutions must equalize access to aid and resources.

The third scenario is, not surprisingly, a modest path that offers an alternative lying between these extremes. This scenario recognizes that despite the end of the Cold War, national behavior with regard to energy issues is still motivated by a quest for power—an enduring characteristic of nations (Krasner, 1985)—coupled with a greater concern for environmental risk than existed prior to the emergence of global climate change as a policy concern.

This scenario suggests that an iterative process of trial-and-error negotiation may take place in a less hostile atmosphere than would have been possible during the Cold War. However, changes in the international balance of power—and in the environmental conditions contributing to global warming—are unlikely to occur as rapidly as changes in the former Soviet Union. I call this scenario "iterative functionalism" because it suggests that nation-states gradually acquire the ability to refine and replicate rules, agreements, and conditions for cooperation and compliance in larger and more complex fora and contexts (Feldman, 1995).

Anderson's major premise is that policymakers in the world's leading financial institutions (e.g., the World Bank, the GEF) are beginning to realize that environmental degradation and its further impact on economic decline and decay are linked to population growth, hunger, and poverty. This being the case, the first scenario above (realistic and pessimistic) makes little long-term sense. It will probably be rejected as developed countries ponder realistic options for responding to climate change. As of the spring of 1995, there is evidence from the first meeting of the Conference of the Parties in Berlin that even those countries most averse to setting specific "targets and timetables" to reduce greenhouse gases (such as OPEC members) have agreed, in principle, to cooperate in trying to forge a workable emissions reduction protocol to the FCCC (Trexler, 1995).

In fact, the United States and OPEC have both announced their support for a protocol development process that after the turn of the century would impose specific milestones for emissions reductions. In addition, U.S. support for the GEF has continued to flourish, and support for greater international energy cooperation is visible in many current aspects of U.S. policy (e.g., the Clinton administration's Climate Change Action Plan).

Despite these developments, however, it is also clear that the second, more "optimistic" scenario is probably far too idealistic without political reform in developing countries. Morgenstern clearly shows why this is the case. Turning to the same negotiations that led to and continue to follow from Rio, developed counties like the United States have maintained a strong reluctance to lend technologies and other forms of assistance to countries that are unprepared to document how they use these new resources for reducing greenhouse gases. They are also reluctant to support regimes that are inclined to pick "winners" and "losers." This has become a much more salient issue since the 1994 U.S. congressional elections.

Also consistent with Morgenstern's analysis, however, the development of a "green" index of economic and energy activities represents one area of international scientific activity currently taking place under the Rio Convention. This effort may, as Morgenstern postulates, lead to greater cooperation among countries. Currently, Rio Convention signatories are trying to account for the potency of different greenhouse gases in an effort to assess more accurately the relative contributions of different activities to global climate change. Annex 1 signatories have also agreed, through the issuance of "national communications" to the FCCC, to undertake inventories of their sources and sinks of greenhouse gases. Such assessments may lead to more eclectic options for abating greenhouse gases (cf. Bohm).

Moreover, these same signatories are also exploring novel methods to reduce greenhouse gases including "joint implementation" (the abatement of greenhouse gases through the sponsorship by developed countries of energy-conserving and energy-efficiency projects in developing countries). These methods rely on markets (as urged by Morgenstern) rather than on command and control regulation. There are currently six joint-implementation demonstration projects taking place, all designed to work by forging an agreement between a developed-country enterprise (e.g., an electric utility) and a host-country enterprise. Developing nations benefit by receiving revenues and technologies, and developed nations benefit by receiving emissions credits. This ability to generate realistic cross-national dialogue is also at the heart of Bohm's account of the post–Cold War energy policy-making environment.

This leaves us with the last scenario, which tries to strike a realistically achievable path given these activities, and given Bohm's argument that the decline of geopolitical rivalries heightens prospects, especially in light of the common consensus that global climate change poses an international risk that requires eclectic solutions. Neither "hard" nor "soft" energy pathways alone can prevent global warming, given the dramatic energy consumption increases forecast for developing countries.

While all of this warrants confidence in the success of international negotiations aimed at reducing atmospheric concentrations of greenhouse gases and the equitable transfer of resources, three things are required to make this path work: (1) acceptance of the fact that international cooperation to reduce greenhouse gases is a learning process that requires access to the best scientific data and to the sharing of the best technologies; (2) a willingness to let the process of negotiation and cooperation shape perceptions rather than the other way around; and (3) a redefinition of *national security*. Bohm firmly believes that the latter is possible in light of the end of the Cold War. An enlarged definition would include ecological security, protection of common goods, and concern for future generations. If Anderson is correct, it is not unreasonable to posit that improvements to the process by which developed and developing countries negotiate transfers of resources could also produce consensus.

REFERENCES

Benedick, Richard E. 1991. *Ozone Diplomacy: New Directions in Safeguarding the Planet.* Cambridge: Harvard University Press.

Bodansky, Daniel. 1995. "Overview: From Rio to Berlin." *Environment* 37 (2): 2–3.

Feldman, David L. 1995. "Iterative Functionalism and Climate Management Organizations: From 'Intergovernmental Panel on Climate Change' to 'Intergovernmental Negotiating Committee.' " Pp. 189–209 in *International Organizations and Environmental Policy,* edited by R. V. Bartlett, P. A. Kurian, and M. Malik. Westport, Conn.: Greenwood Press.

Kinley, Richard. 1995. *Governments Act on Treaty Commitments: A First Review. United Nations Climate Change Bulletin* 6. Chatelaine, Switzerland: Interim Secretariat for the UN Climate Change Convention.

Krasner, Stephen D. 1985. *Structural Conflict: The Third World against Global Liberalism.* Berkeley: University of California Press.

Mitchell, Ronald B. 1994. *International Oil Pollution at Sea: Environmental Policy and Treaty Compliance.* Cambridge: MIT Press.

Trexler, Mark C. 1995. *The Mitigation Monitor: A Client Service of Trexler and Associates.* Oak Grove, Ore.: Trexler and Associates, April.

United Nations Climate Change Bulletin. 1995. Issue 8. Chatelaine, Switzerland: United Nations Environment Program.

11. Energy, Environment, and Economy

Complementarity and Conflict in the
Search for Sustainable Growth

DENNIS ANDERSON

Can the world's rising demands for commercial energy be met *and* pollution from energy production and use be reduced to socially satisfactory levels? Barring calamities, and in an energy-efficient scenario, aggregate commercial energy demand will likely rise two- to threefold over the next thirty years owing to the growing weight of developing countries in the world economy. If that happens, energy demand would rise from 8 billion tons of oil equivalent today to around 20 billion tons by 2025. For electricity, whose share of final energy markets is rising, worldwide installed capacity would rise from roughly 3 million to 7–10 million megawatts (MW), or five to ten times aggregate installed capacity in the United States today. These would be extraordinary expansions, and even then the world's demand would probably not be at its peak because of growing population and because per capita consumption levels in developing countries would still be only one-fifth the levels in today's high-income countries.

Achieving what has come to be called sustainable development requires meeting the growing energy demands of developing countries. Energy efficiency is often appealed to as a means of addressing energy pollution problems. Efficiency is certainly important for many reasons. However, because developing countries are growing, energy efficiency alone will not solve environmental problems. Historically, energy efficiency has been a source of growth, not contraction, in the energy industry. A situation in which the per capita supply and demand for commercial energy remain low in developing countries is one in which billions of people remain dependent on fuel wood and dung for cooking and are without electricity for lighting, motive power, and many other purposes. Such a situation would be environmentally unsustainable, socially indefensible, and ultimately self-defeating because it would be associated with high population growth and a failure of developing countries to eliminate poverty and attain prosperity.

The first section of this chapter develops these points further, and develops energy demand scenarios under more buoyant assumptions. Subsequent sections argue that the key to reducing pollution is the development and use of low-polluting technologies and practices. In the near term, "clean-coal" technologies, natural gas, reformulated fuels, and emissions controls on vehicles, and in the longer term, renewable energy technologies, hold much promise for reconciling large increases in energy production and use with high levels of local, regional, and, eventually, global pollution abatement.

World Energy Demand

Levels of per capita commercial energy consumption in developing countries are low when compared with those in industrialized countries. Per capita commercial energy consumption in the United States is 8 billion tons of oil equivalent per year, or eighty times greater than in Africa, forty times greater than in South Asia, fifteen times greater than in East Asia (excluding Japan), and eight times greater than in Latin America (see fig. 11.1). Similarly, electricity consumption is roughly 13,000 kilowatt-hours (kWh) per capita in the United States as compared with around 600 kWh per capita in developing countries. If, as incomes grow, people in developing regions come to consume about a quarter of the electricity used by the average U.S. citizen, even allowing for population growth, total world consumption would rise to around 30 billion tons of oil-equivalent energy. Meanwhile, electricity capacity would rise to over 30 terawatts (TW). These figures are, respectively, four and ten times today's demands for energy and electric power. Such elementary arithmetic immediately suggests why large demands lie ahead, assuming continued economic progress.

Reasons for Increases in World Energy Demand

More formally, there are three reasons to expect a major increase in world energy demand: population growth, income growth, and substitution of commercial for traditional fuels such as dung and fuel wood in developing countries.

Population and Income Growth

Current "base-case" projections of world population growth by the World Bank and the United Nations show an increase from 5.2 billion in

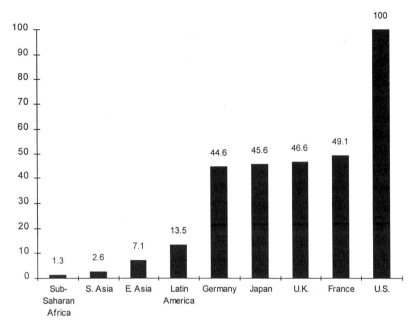

Figure 11.1   **Per Capita Consumption of Commercial Energy Relative to per Capita Consumption in the United States, 1990 (U.S.A. = 100)**
Source: World Bank, 1992.

1990 to 8.3 billion by 2020 and to nearly 10 billion by 2050, with a stationary size of about 12 billion (World Bank, 1992). A higher stationary population of 23 billion, assuming slower fertility decline, has also been estimated, as has a lower case of 10 billion, which assumes more rapid fertility decline. More recently Bongaarts (1994) has suggested a fourth scenario, which estimates that stationary population could be as low as 7.5 billion with more ambitious family planning. These projections ignore the possibility of epidemics. The lower-population-growth scenarios also (rightly) assume that an increase in life expectancy due to better health is a desirable end of development.

Analysis of the differences among the four cases reveals why it is a situation of economic success combined with good social policies that is more likely to produce a low-population scenario (and why the high-population scenario, in contrast, would likely arise only if there were major failures in economic and social policy).

The higher the level of per capita incomes, and the more development succeeds in reducing poverty, the lower the ultimate stationary population

(Birdsall et al., 1984; Bongaarts, 1994: 771–75). Higher per capita income and reduction of poverty are associated with (among other things) more education, greater availability of family-planning services, more social security, and a desire by families to increase the time they allocate to directly remunerative as opposed to household work. All these reduce fertility rates. Reductions in infant mortality also reduce fertility rates because of greater family stability and security (not to say relief from despair). As Bongaarts remarks, "No population in the developing world has experienced a sustained fertility reduction without first having gone through a major decline in infant and child mortality" (Bongaarts, 1994: 774).

Family planning and education will also affect population growth and the eventual size of the stationary population. The difference between Bongaarts's scenario and the low-population scenarios of the World Bank and the United Nations is that Bongaarts assumes even more successful family planning and education programs. He proposes three policies for reducing population growth rates: reducing unwanted pregnancies by strengthening family-planning programs; reducing the demand for large families through investments in health and in education for both children and adults; and addressing "population momentum" by encouraging women to have children later, something that he and others believe could be achieved through greater investment in secondary education for girls.

These connections among population growth, economic development, and social policy have an implication that is often overlooked in discussions of energy policies and development: it concerns the contribution that economically efficient energy pricing policies can make indirectly to health, family-planning, and education policies in developing countries by reducing budgetary stresses.

Market-oriented economies have comparatively minor deformities in their pricing systems, and the benefits of energy production and use are fairly well reflected in economic returns on investment. But in economies where price deformities are large (e.g., many developing countries and the economies of Eastern Europe and Central Asia), the side benefits of price reforms are considerable. Energy subsidies are a major hemorrhage on public revenue and, indirectly, on socially desirable public programs. The 1992 *World Development Report* estimated that the annual *incremental* cost of putting developing countries onto a sustainable growth path would amount to roughly 2–3 percent of GNP, or roughly $100–150 billion per year at their present income levels (World Bank, 1992: chap. 9, table 9.1).

These costs include public investments in soil conservation ($15–20 billion per year), controlling a wide range of emissions and effluents from industry and transport ($15 billion per year), additional agricultural research and development ($5 billion per year), expanded primary and secondary education for girls (a much neglected area), and expanded family-planning programs (an increase of $7 billion per year over present levels of $4–5 billion).

Gains in public revenue from price reforms in the electricity sector alone in developing countries could finance such programs virtually in their entirety. According to a recent survey, electricity prices in developing countries averaged only 4 cents per kWh in 1991, less than half of the marginal cost of supply (Heidarian and Wu, 1994), while revenue shortfalls approached $120 billion per year, with much of the loss being borne in various ways through government subsidies. (Energy subsidies thus amount to twenty times the costs of the most ambitious extensions to these countries' family-planning programs.) Improving electricity pricing is of course fundamental to economic and energy efficiency; but it is also important to reduce budgetary stresses in developing countries and to make the electricity industry financially self-sufficient. This is not to suggest that the financial gains from price reforms in the electricity industry, or from eliminating unnecessary losses to public revenues, should all be used to finance these programs. The calculations merely illustrate that substantial rewards can accrue to development, arising from *economic* efficiency in energy production and use.

It follows from the above that any scenario consistent with the broad aims of economic development needs to be based on three assumptions:
—continued economic growth in developing regions (and a significant increase in currently slow-growing regions)
—success in family-planning and education programs to improve human welfare and attain lower population levels
—achieving *economic* efficiency in energy production and use, especially price and cost efficiency

Alternatively, one could work with scenarios of low economic growth and a failure to contain population growth, and postulate various kinds of calamities. But this would beg the question of how to develop energy and environmental policies that would enable developing countries to achieve and sustain economic growth. Lower rates of economic growth do not necessarily lead to lower energy demands and environmental stresses. In fact, the opposite is true. Low income growth would likely mean higher population growth, and an intolerable rise in the populations dependent on dung, crop residues, and fuel wood (biofuels) for cooking.

Substituting for "Traditional" Biofuels

Over 2 billion people in developing countries, mostly in rural areas and towns, depend on biofuels—crop residues, animal dung, wood and charcoal—for cooking. In addition, 2 billion people have no access to electricity. Biofuels are the largest single source of final energy consumption in most countries, accounting for over 30 percent of overall supplies on average (final electricity consumption is less than 20 percent) and over 80 percent in Africa, and their use is a major source of environmental damage. They give rise to high levels of indoor air pollution and are damaging to health (Smith, 1988). The diversion of dung and crop residues from the land to the fireplace depletes soil productivity and is considered a significant threat to agricultural sustainability and human health. Fuel-wood consumption and conversion to charcoal for cooking in urban areas is a major source of deforestation in developing countries.

Finally, cooking with biofuels is perhaps the single largest source of energy inefficiency in developing countries, and of CO_2 emissions (as well as local pollutants). Biofuels are fully four times less efficient than gas or kerosene as a cooking fuel, unless used in conjunction with high-efficiency cookstoves, and their associated emissions, particulate matter, and sulfurous compounds are several orders of magnitude higher.

Taking the base-case scenario of population growth, and without further expansion of the supplies of modern energy forms—gas, liquid fuels, electricity, and renewable energy—over 6 billion people would become dependent on wood and dung for fuels over the next four decades, and 6 billion people would also be without electricity. It is difficult to estimate whether the supplies of wood, crop residues, and dung could support such a colossal population. But it is reasonable to conclude that this situation would be socially and environmentally unsustainable, and possibly calamitous. Any policy aiming to sustain development must therefore include a long-term transition to commercial fuel use—including renewable energy—for the huge populations still without them.

The most important factor influencing the transition to modern fuels is per capita income growth. There is a correlation between income and share of traditional and modern fuels in total consumption (fig. 11.2). There is also evidence that the transition in a high-growth economy can be much

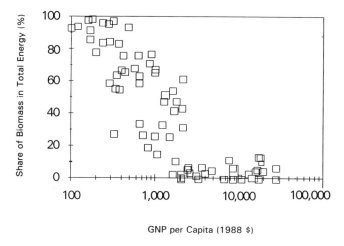

GNP per Capita (1988 $)

Figure 11.2 Traditional Energy versus Income, Selected Countries, 1988
Source: World Bank data.

quicker than it historically has been in the industrial countries. For example, the shift from fuel wood to modern fuels—coal, gas, oil, and electricity—took sixty years in the United States, while in Korea the same transition took about thirty years (Pearson, 1988).

Supply policies of the energy industry also influence the rate of transition. Over the past twenty years, nearly 1 billion people in developing countries, or four times the population of the United States, have become newly connected to electricity grids. While this is an immense achievement, it has not alleviated the fuel-wood problem, since electricity is rarely used for cooking in rural areas even when it is available. It is, however, symptomatic of attempts by the energy industry to expand supplies to a larger share of growing populations.[1] In addition, it represents a major gain in energy efficiency for at least one use of energy—lighting—because the incandescent lamp is roughly one hundred times more energy-efficient (measured in lumens per watt) than the kerosene wick lamp it displaces (van der Plas, 1994a).

Assuming good income growth and ambitious investment policies, the transition to modern fuels is going to take several decades in low-income countries, because of population growth and the huge number of people still unserved. There, energy policy needs to be concerned not only with the expanded use of modern fuels but also with environmentally satisfactory ways of providing biofuels for cooking and other uses in low-income countries that remain dependent on these resources (Gregerson, Draper, and Elz, 1989; Barnes et al., 1993). As regards biofuels, there are broadly three elements in the policies required.

First, there needs to be an expansion of participatory approaches to afforestation on farmlands and woodlands based on social forestry (Gregerson, Draper, and Elz, 1989). Planting trees and shrubs in association with agriculture accomplishes:

—Higher planting rates. Farmers outnumber foresters by several thousand to one. Involving them in afforestation programs makes possible a significant acceleration of planting rates and, thus, an increased production of biomass.

—Reduced soil erosion and increased crop yields due to the favorable effects of social forestry practices on microclimates, the retention of moisture in soils, and the supply of soil nutrients.

—An increase in fuel wood, fodder, mulch, and other by-products of local trees and shrubs.

Afforestation is therefore an excellent area for public investment—to broaden the messages and services of agricultural education and extension, and to encourage the establishment of nurseries and supporting afforestation services. It is capable of generating good economic rates of return (Anderson, 1989).

Second, investments in small-scale applications of renewable energy (e.g., photovoltaics for lighting, solar cooking, renewable energy in agriculture and agro-industries) are often merited. Renewable energy sources are increasingly becoming a cost-effective commercial energy source for rural areas.

Third, support for wood-stove programs is required. Although initial efforts in many countries over the past fifteen years have been disappointing, there seems no option but to persevere and build on successful examples, of which there are several (Barnes et al., 1993).

Energy Demand Scenarios

It is difficult to construct a scenario in which world population growth is kept manageable, developing countries prosper, and commercial energy consumption does not grow. Everything points to very large increases in the world's demand for commercial energy. As discussed, the scenarios of low demand relative to today's levels are socially and environmentally disquieting because they imply huge populations dependent on fuel wood and dung

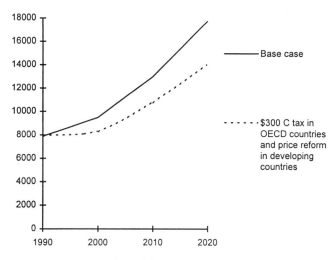

Figure 11.3   Two Scenarios of World Commercial Energy Demand, 1990–2020 (millions of tons of oil equivalent)
Source: Anderson and Gammons, 1993.

for cooking, and are pessimistic about both income growth and, by implication, population growth.

Figure 11.3 projects commercial energy demands that typically emerge under the assumptions discussed above. A recent study by the International Energy Agency (IEA) has obtained similar results (OECD/IEA, 1993). The economic assumptions are as follows:

—Continued economic growth in Asia, continued recovery and growth in Latin America, and recovery and achievement of growth in Africa. Markets are assumed to have "matured" in OECD countries, and to decline somewhat with economic and efficiency reforms in Eastern Europe and Central Asia.

—The base-case population projections of the World Bank and United Nations.

—Continued substitution of commercial fuels for fuel wood, dung, and crop residues for cooking and small-industry uses.

To assess the effects of possible gains in energy efficiency on demand, a second scenario in figure 11.3 shows the effects on demand of hypothesized increases in energy taxes and prices. The intention is *not* to suggest that such taxes are desirable—a quite different policy for reducing carbon emissions is discussed later—but to show that even radical policies to curtail energy

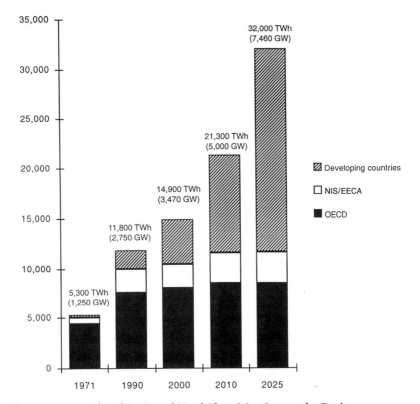

Figure 11.4   **Actual and Projected Total Electricity Output, by Region, 1971–2025 (terawatt-hours per year)**
Source: Estimates are based on the assumptions discussed in the text.
Note: Numbers denote total electricity output per year (TWh) and net installed generating plant capacity (GW). NIS/EECA = newly industrialized states/Eastern Europe and Central Asia.

demand growth have only secondary effects on world energy usage. The second scenario illustrated in figure 11.3 estimates the likely effect on world energy demand of carbon taxes of $300 per ton being gradually introduced in OECD countries, and of price reforms in developing countries that remove all subsidies on energy production and use and gradually raise energy taxes to the high levels obtaining in Europe today for gasoline and diesel fuels. (The underlying assumptions regarding income and price elasticity and prices are summarized in Appendix B of this volume.) The IEA earlier undertook assessments of the effects of carbon taxes on demand, with practically the same results obtained here (OECD/IEA, 1993).

Figure 11.4 shows some recent projections of electricity demand for an "energy-efficient" case. It is not derived from any formal model but is based on assumptions similar to those above: successful achievement of economic growth in the developing countries, electricity prices gradually rising to reflect the costs of supply, and a rising share of population being supplied with electricity. Figure 11.5 shows the assumptions about average per capita consumption levels. Even with very large electricity demand increases in developing countries, per capita consumption would still be very small relative to that in OECD countries, and barely one-fifth of that in the United States today.

Can there be further savings from energy efficiency? The above scenarios focus mainly on the role of market-based incentives to achieve efficiency. But a rather different, "bottom-up" assessment by Lee Schipper of energy use and prospective demands in thirteen developing countries (Schipper, 1993) arrives at much the same conclusion. Under the "efficiency-push" and "vigorous-effort" scenarios, he estimates the likely energy demands when highly efficient end-use technologies are incorporated into the capital stock. (The preceding analysis implicitly assumes that such technologies would be encouraged through the pricing system.) Schipper also finds that developing-country demands would probably rise by a factor of three to four over the next three decades, and by much more in an energy-inefficient scenario.

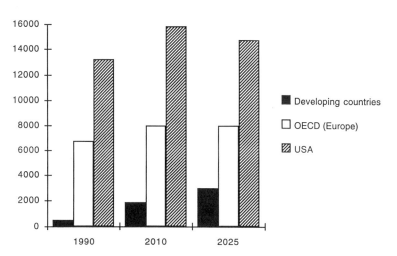

Figure 11.5   **Actual and Projected per Capita Electricity Consumption, by Region, 1990–2025 (kilowatt-hours per year)**
Source: Estimates are based on the assumptions discussed in the text.

Furthermore, energy efficiency has historically often been a source of growth in the industry because it reduces costs and prices and increases the benefits of energy consumption (Anderson, 1993a). For example, the price of electricity has fallen by a factor of fifteen or more in the United States this century, largely because of greatly improved power-station thermal efficiencies, which have risen from around 5 percent in 1900 to 45 percent and greater for new stations today. Assuming a price elasticity of -0.5, this alone would have led to a fourfold increase of demand. Similarly, demand-side energy-saving innovations have increased rather than decreased applications of energy, and thus have increased demand (e.g., the fluorescent lamp invented for commercial lighting in the 1930s). While energy efficiency does not always lead to reduced demand, it still has significant economic benefits.

Thus, even with efficient production and use, the levels of world energy demand and pollution are likely to rise appreciably without changes in energy supply technologies and policies.

Energy Supply Technologies and Policies

Reducing Local Pollution

In the industrial countries, the main path to reducing local pollution has historically been development and use of low-polluting technologies and practices. Until the middle of the present century, the principal environmental concern associated with energy production and use was urban smog from particulate-matter (PM) emissions from power stations, factories, and the use of coal as a domestic heating fuel. Over a period of less than two decades, the problem was virtually solved by the use of PM controls in power stations and factories and the use of gas, electricity, and, in some instances, "clean-burning" coal and coke as a domestic fuel. The first two rows of table 11.1 indicate the extent of abatement of PM emissions from such substitutions. Remarkably, energy use could in theory expand several hundredfold from today's levels yet produce *less* pollution from PM from the above-noted sources than occurred in the past. The importance of technical progress and substitution in resolving the apparent conflict between energy use and the environment cannot be overstated.

The next two rows of table 11.1 show the percentage abatement of the technologies developed to address problems emerging between the 1960s and the 1980s once the PM problem was addressed: acid deposition from SO_2 and NOx, photochemical smog from the action of sunlight on NOx, and lead poisoning from the use of lead as an octane enhancer in gasoline.

Table 11.1.   Pollution Abatement from Pollution Control Technologies

Pollutant or Source of Damage	Percentage Abatement	Means of Abatement
PM from power stations and factories	99.9	Electrostatic precipitators, baghouse filters
Smog from PM (home fires)	>99	Natural gas/"clean" coal
SO_2 and NOx from power stations	90–95	Scrubbers, fluidized bed combustion, "lo-NOx" combustion and catalytic devices
Leaded fuels, CO, VOCs, and NOx from car exhausts	99 (100 for lead)	Unleaded fuels, catalytic converters, reformulated gasoline
Indoor air pollution from wood fuels	>99	Use of commercial fuels in homes
Soil erosion, nutrient loss from use of firewood, dung, and crop residues for cooking in developing areas	>99	Agro-forestry, vetiver grasses, contour cultivation, terracing, and use of commercial fuels in homes
Global warming from CO_2	100	Renewable energy

Sources: World Bank, 1992, and supporting background papers.

In each case there has been a technological response that enabled local pollution to be reduced in industrial countries. In some instances (e.g., for NOx) the full benefits have yet to be felt in industrial countries because new technologies have not been fully incorporated into the capital stock. However, there is some evidence that these technologies are beginning to reduce atmospheric pollution significantly (OECD, 1991).

In developing countries there is still little use of low-polluting technologies, and many countries seem to be repeating rather than avoiding the worst excesses that cities in the industrial countries once experienced. It is beyond the scope of this chapter to review the extent of local pollution from energy production and use. Such a review would in any case merely duplicate a large number of excellent country and international studies already available.[2]

There is a broad range of technologies and practices for developing countries to draw on, and not only are these approaches proven, but the costs, with exceptions, are small relative to those of energy production and use.[3] For *road transport,* for example, the costs of the emissions controls described in table 11.1 amount to about 15 cents a gallon (World Bank, 1992).

This estimate includes the extra cost of the fuels plus annualized capital costs of tailpipe controls on vehicles divided by the average miles traveled per vehicle per year. Such costs are not trivial in absolute terms, and when environmental policy decisions are being made, they need to be weighed against the benefits of pollution abatement. Nevertheless, they are a small percentage of the costs of road transport.

For *electric power* the costs of PM controls amount to approximately 2 percent of electricity supply costs, and for PM, SO_2, and NOx together, approximately 5–10 percent for "end-of-pipe" controls on conventional coal boilers. On the other hand, new fluidized bed combustion technologies and "topping" cycles[4] hold the promise of greater efficiencies and lower costs. If gas is available for combined-cycle power plants, costs are lower compared with the plants of the past that used stoker or pulverized fuel (coal) boilers, again because of improvements in thermal efficiencies.

The costs of pollution abatement are also small when compared with the economic benefits from "win-win" economic efficiency reforms. The financial losses arising from the subsidy of electricity generation in developing countries amount to $120 billion per year (at the time prices were last surveyed). This amounts to fifteen times the total capital expenditure required to abate all local pollution from coal-fired plants to socially satisfactory levels using "clean-coal" combustion technologies (World Bank, 1992). Pollution would be reduced directly through the energy efficiency induced by higher prices (by about 30 percent, assuming a price elasticity of -0.5). There would also be a sizable economic gain beyond the benefits of pollution abatement, and a sizable financial gain after allowing for pollution control costs. The win-win possibilities in land transport are also potentially large (e.g., through electronic tolling and congestion pricing in urban areas) (Anderson and Cavendish, 1992).

Table 11.1 also indicates the possibilities for reducing damage to public health and the rural environment through the use of biofuels. The use of commercial energy (or of improved cookstoves) would practically eliminate indoor air pollution, one of the worst environmental problems in developing countries today. We also know that there are practices for securing a better supply of biofuels while actually *enhancing* the productivity of soils, practices that improve moisture retention and nutrient supply while reducing erosion. One such practice is social (or agro-) forestry. Others include the use of vetiver grasses, contour cultivation, and terracing on sloping lands (Doolette and Magrath, 1990). Aside from reducing soil erosion and helping to make agriculture sustainable, such practices can raise yields by 20 percent or more and have good economic returns to invest-ment. Once again we have an example of an environmental and health problem capable of being solved while outputs (e.g., agriculture and energy supplies) are improved, not worsened.

This summary of some practices for addressing local environmental problems arising from energy production and use in developing countries shows that the problems are not fundamentally technical, managerial, eco-nomic, or financial in nature. On all counts, options are available and, if used, would greatly improve the quality of the environment and public health. In some cases (e.g., NOx or PM from diesels) there remains much to do to develop the technologies and practices further. There are also en-vironmental problems emerging from the use of MTBE (methyl tertiary butyl ether) as an octane enhancer. However, the biggest jobs are in public education and policymaking.

Simulating the Effects of Policy Alternatives

Figure 11.6 illustrates the kind of results that typically emerge from simula-tions of alternative policies. It specifically considers the case of electric power in developing countries, and has been previously published (World Bank, 1992). The upper curve shows the likely growth of pollution with no change in practice (pollution is indexed, with emissions in 1990 equal to

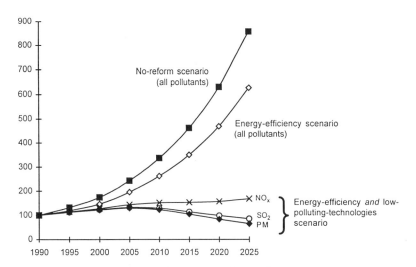

Figure 11.6 **Three Scenarios of Emissions from Electricity Generation in De-veloping Countries, 1990–2025 (index, 1990 = 100)**
Source: Anderson and Cavendish, 1992.

100). The next curve downward shows the likely growth if economic and other reforms to improve efficiency are gradually phased in. For reasons previously discussed, efficiency helps somewhat, but output and emissions are still expected to grow exponentially because of growing demand in developing countries. The decisive impact on pollution abatement comes from the switch to low-polluting technologies, as shown by the lower curves. Pollution continues to rise initially in the latter case before it declines; this is due to growth in demand and the time needed to incorporate new low-polluting technologies and practices into the capital stock.

Similar results can be derived for other classes of pollution problems. The key variables are growth of demand and output; the level of emissions per unit output of polluting relative to nonpolluting technologies and practices; and the rate of adoption of low-polluting practices. The process can be readily modeled as follows.

The rate of introduction of low-polluting practices and technologies depends on the size of the capital stock; how well developed the technologies are; the retirement rates of existing equipment; whether retrofitting is feasible; and costs (and, thus, the strength of the incentives provided by environmental taxes and regulations). At the simplest level, the model takes the form:

$$E_t = [(S_t)c + (1 - S_t)p] [Q_t/(p \cdot Q_0)]$$

where E_t represents total emissions in year t relative to the base year; S_t the share of low-polluting practices in the capital stock, assuming $S = 0$ initially; c the emissions coefficient per unit output of these practices; p the emissions coefficient for the polluting practices; and Q_t and Q_0 total output in periods t and 0 respectively. Once substitution to low-polluting practices is completed ($S = 1$), emissions will be reduced according to the ratio c/p times the ratio by which output has increased. The data in table 11.1 show that for energy production and use, and given wise choices of technologies and practices, values of c/p can be achieved that are one or more orders of magnitude lower than the prospective increases in energy demands discussed earlier.

Against this, the time required for decision makers to introduce new policies and for producers to respond can be quite long for some pollutants. The time pattern of substitution typically follows an **S** or logistic curve and can be calculated using "vintage" models of growth. For PM, for example, the lags were relatively short, once the countries introduced the required policies. But in the United Kingdom it took more than a century, and repeated crises and public outcries against "killer smogs," before such policies were introduced (Brimblecombe, 1987). The lags were also short for introduction of unleaded fuels in the industrial countries, once the problem was identified. But the lags have been quite long for addressing the problem of acid deposition, on account of the greater difficulties and costs of developing low-emissions alternatives. Lags are also likely to be long for the introduction of social (or agro-) forestry in rural areas and for the adoption of improved methods of cultivation, since a significant effort is required to encourage farmers to change practices through agricultural education, research, and extension. Land or property rights reforms may be necessary in many areas, and the absence, as well as achievement, of such reforms can present formidable difficulties. The available simulation studies of technical responses to global warming also point to long lead times and lags, typically half a century or more before S reaches 0.5 (see Grubb et al., 1993).

Global Warming and the Case for a Solar Initiative

If it becomes necessary to restrict the use of fossil fuels on account of global warming, will the non-net-carbon-emitting or backstop technologies become available to meet the world's demands for commercial energy?

Until recently, discussions of this question proceeded as if there were no economically or operationally feasible substitutes for fossil fuels other than nuclear power. However, in recent years a range of renewable energy technologies for both small- and large-scale supplies have emerged: photovoltaics (PVs), for which there are more than fifty approaches being developed and tested based on alternative materials and cell designs; thermal solar schemes; wind energy; and advanced biomass power generation concepts. High oil prices in the period 1973–85 stimulated much commercial research into these technologies, but economic interest waned briefly following the collapse of oil prices in 1986 and was kept alive only by a few research laboratories and industrial pioneers. Continued progress and positive operational experience together with concerns about global warming are now leading to a revival of interest. It is remarkable that the costs of some renewable energy technologies (PVs, thermal solar, and wind) have fallen by more than oil prices over the past ten to fifteen years. Public policies have yet to catch up with these developments. Some recent developments, their economic importance, and needed policies are summarized below.

Costs and Operational Performance of Solar Technologies

Recent developments and cost reductions in solar energy technologies have been substantial (Johansson et al., 1992; Ahmed, 1994). Consider, for example, photovoltaics, for which historical and projected costs (in 1990 prices) are shown in figure 11.7. In the early 1970s, costs of PV modules were several hundred thousand dollars per kWp (peak kilowatt capacity), and applications were limited to aerospace programs and other specialized uses. By around 1980, costs had fallen tenfold, to the range of $25,000–50,000 per kWp. By 1990, costs had fallen to $6,000 per kWp, and PVs had become commercially viable and technically proven for a wide range of small-scale applications (e.g., village and domestic lighting, water pumping, battery charging, and supplies to rural health clinics) in developing countries.

There is also a growing luxury market in the industrial countries, where experiments with PVs as a source of supplementary grid power are now being conducted. The growth applications are well illustrated in a recent report by Robert van der Plas, who notes that in Kenya more rural households were newly supplied with electricity from PVs over the past five years than were newly supplied from the grid (van der Plas, 1994b). Interestingly, this has been a purely private market development, unaided—and in fact discouraged—by Kenya's tax policies. The costs of PV modules are now quoted in the $4,000–6,000 range ($8,000–10,000 including balance-of-system costs), and further progress is expected because of:

—Scale economies and technical progress in production. World output has grown from around 1 MW per year fifteen years ago to over 60 MW today, a growth rate of over 30 percent per year, albeit from a small base. This is still a small market, but the technologies are modular, and economies of scale and the technical possibilities for batch production have barely been exploited.

—Further developments in cell, module, and systems design and improvements in conversion efficiencies. Materials development, the use of multijunction devices and novel cell designs to capture a higher proportion of the solar spectrum, and the use of concentrators (Fresnel lenses) to focus the sunlight onto high-efficiency cells are further areas of rapid development.

There are convincing engineering and economic reasons for expecting further progress, and the U.S. Department of Energy projects that with market expansion, costs should eventually decline to around $2,000 per kWp (including balance-of-system costs), a plausible projection supported

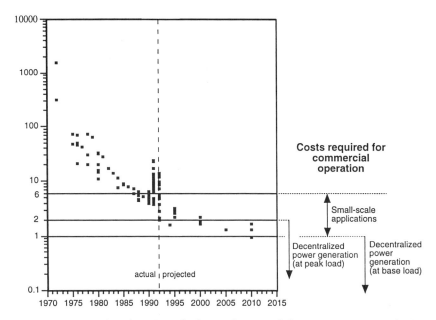

Figure 11.7 Actual and Projected Photovoltaic Module Costs, 1970–2015 (U.S. dollars per peak kilowatt capacity, 1990 prices, logarithmic scale)
Source: Ahmed, 1994.
Notes: The range of costs marked on the graph show the photovoltaic module costs required to compete with small-scale applications and with decentralized power generation (assuming supply costs of 8–10 cents/kWh at base load and 16.45 cents/kWh at peak load, which include generation, transmission, and distribution). The ranges are approximate and assume that balance-of-system costs will come down commensurately with module costs. (Balance-of-system costs are not shown but are assumed to be roughly the same as module costs.)

All module costs for years up to and including 1992 are actual; those after 1992 are projected. The "spread" in the points reflects the spread in costs of different technologies, which are at different stages of development. The size of the module used also affects cost, as does the size of the order.

by learning-curve studies modeled on data similar to that provided in figure 11.7. If this were to happen, PVs would become economical for grid-connected applications in the distribution networks of countries with good solar insolations, and also for the emergence of independent or "distributed" utilities.

Progress in thermal solar schemes is also noteworthy. Parabolic trough technologies have been proven for large-scale power generation, with costs of $3,000 per kW and 12–17 U.S. cents per kWh (fig. 11.8). Steam conditions compare favorably with those provided by fossil and nuclear stations, typi-

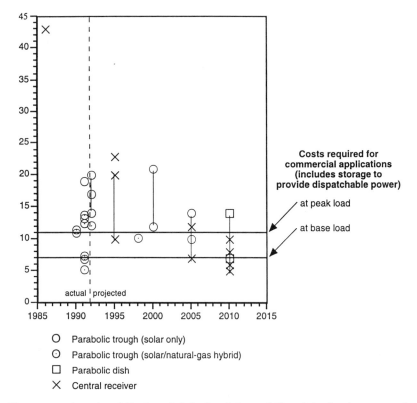

Figure 11.8   Actual and Projected Calculated Cost of Electricity from Large-Scale Solar-Thermal Technologies, 1985–2015 (U.S. cents per kilowatt-hour, 1990 prices)
Source: Ahmed, 1994.
Note: Costs for years up to and including 1992 are based on the technology of the time, and are data from actual plants as well as results of engineering studies; costs for years after 1992 are from projected data.

cally 1,000 pounds per square inch and 700 degrees Fahrenheit, while the availability of the solar fields in the Kramer Junction Plants in California has risen to 99 percent. Costs are still high relative to fossil-fired power stations, but they are appreciably lower than the costs of nuclear power plants commissioned in the United States in the 1980s. They also compare favorably with the costs of some hydro schemes in developing countries. Further, as with PVs, scale economies in manufacture and technical possibilities have barely been exploited. Experience with solar-thermal power stations dates back only to the mid-1980s, with only 350 MW having been built.

Research on a wide variety of materials and design concepts in solar energy, both on PVs and in thermal solar, has proved to be fertile. There is ample scope for additional gains in conversion efficiencies, from the present range of 7–15 percent for field applications to the 20–30 percent range, the latter having already been achieved in laboratories. The potential is especially large in developing countries where solar insolations are high and energy markets are growing rapidly. Significant progress has also been made in the use of wind and biomass resources for power generation (Johansson et al., 1992).

Short lead times are another notable feature of PVs, thermal solar schemes, and wind power. Construction time for California's solar thermal plant was nine months, while PV systems can be installed in even shorter times. Times quoted for wind are comparable to those for thermal solar. Only biomass-fired power generation projects have long lead times among the solar energy technologies, on account of the need to grow the biomass feedstocks. However, even here there is promise of shorter lead times using biomass from annual crops rather than wood. Another feature of solar technology is the relative ease and speed of decommissioning once a plant has completed its operational life; for all practical purposes, it is a "reversible" technology. Finally, the solar plant can often be maintained while operating, owing to its modularity. This should improve operational performance and reduce maintenance costs.

Abundance of the Solar Resource and Land Availability

Although the point has been repeatedly made, the abundance of the solar resource remains insufficiently recognized. It is a diffuse source of energy, but the areas required to harness it are not large. There are several ways of making the point. One is to compare the land requirements with those of hydro schemes, as shown in figure 11.9. Except for run-of-river schemes and for a few favorable high head sites in deep gorges, the land requirements of hydro range from ten to several hundred times those of solar schemes at today's conversion efficiencies, averaging around twenty-five to fifty times. Solar schemes are capable of supplying, in theory, *five to ten times* the total electricity demands of developing countries today while occupying land areas less than those of current hydroelectric schemes. Whereas a country like Mali, for example, has a hydroelectric potential of 1,500 MW, its solar potential is one hundred thousand times this amount. Similar remarks can be made of many other countries. Moreover, as the World Bank notes in its 1992 *World Development Report* (World Bank, 1992), larger-scale solar schemes are often best located in arid areas with low population density. There is much flexibility in the choice of site. Small-scale PV systems, on the other hand, can be sited unobtrusively on rooftops or on small sites near substations. Hence, solar energy does not have the problems that so often

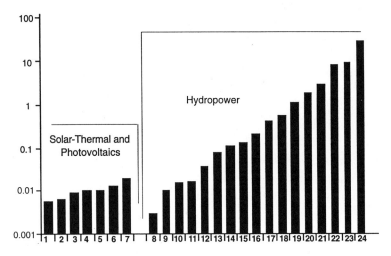

Project (see key below)

Solar-Thermal and PVs: 1. Eurelios central receiver plant; 2. Solar One central receiver; 3. PV concentrator scheme; 4. CESA-1 central receiver; 5. PV concentrator scheme; 6. PV concentrator scheme; 7. Luz parabolic trough.

Hydropower. 8. Nathpa Jhakri, India; 9. Marsyangdi, Nepal; 10. Berke, Turkey; 11. Xingo, Brazil; 12. Kulekhani, Nepal; 13. Zimapan, Mexico; 14. Itapu, Brazil; 15. Aguamilpa, Mexico; 16. Machinadinho, Brazil; 17. Daguangba, China; 18. Yacyreta, Argentina; 19. Tres Irmáos, Brazil; 20. Aswan, Egypt; 21. Samuel, Brazil; 22. Akosombo, Ghana; 23. Balbina, Brazil; 24. Nangbeto, Togo.

Figure 11.9 **Collector Areas of Solar Plants (Photovoltaics and Solar-Thermal) Compared with Inundated Areas of Hydroplants (ratio of square kilometers to megawatts, logarithmic scale)** *Source:* Anderson and Ahmed, 1995.

beset hydro schemes: the competition with agriculture for land and the need to resettle large numbers of families.[5]

Another way of looking at land requirements is to compare them with those for agriculture. Consider the demand scenarios for developing countries. If their aggregate electricity demand rose to 5 million MW over the next thirty years or so, then in theory—assuming present-day conversion efficiencies—their electric power and energy demands could be met by solar schemes occupying an area equal to only 1 percent of the area presently under crops and 0.25 percent of the area under crops and pasture. This would equal 80,000 square kilometers (km) as compared with 8.5 million square km under crops and 30 million square km under crops and pasture. Such hypothetical calculations are intended only to illustrate the point that because of technical developments over the past two decades, technologies are emerging to harness a virtually inexhaustible resource.

A Promising Approach to Addressing Global Warming

These calculations and the technical progress made thus far show that solar energy is emerging as a promising approach to the global-warming problem, should the need arise, and for this reason its development and use should be given greater attention by the Global Environment Facility managed by the World Bank, the United Nations Environment Program, and the United Nations Development Program.[6] The same cannot be said of any other backstop technology at present. After more than forty years of development, nuclear power has not reached the point where it can be confidently recommended to developing countries: costs are still high relative to fossil fuels; construction times have risen to disconcerting levels (over ten years in the United Kingdom and the United States, six to seven years in France and Japan); the scientific, engineering, and managerial requirements are formidable; and, especially important, many unresolved problems remain concerning reactor safety, waste disposal, and decommissioning (Cochran, Morris, and Suokko, 1993). Until such problems are resolved, there is little prospect for new nuclear power investments attracting finance from international institutions or the Global Environment Facility. The exact opposite is true for solar power.

The Components of a Solar Initiative

Despite the impressive gains that have been made, solar power is not commercial, except for small scale "off-grid" applications of PVs, some wind energy projects, and co-generation from biomass residues. Costs are higher than the estimated costs of supplying peak and off-peak power for grid supplies from advanced fossil-fired stations, as shown in figures 11.7 and 11.8. Were it not for concerns about global warming, a sufficient rate of progress with solar technologies might be achieved over the next two decades through a gradual development of markets, modest increases in R&D funding, and tax incentives to encourage further application. As the U.S. Department of Energy has projected, this would probably be sufficient to see commercial applications gradually emerge over the next two to three decades. Because global warming is a public concern, however—over 150 countries are signatories to the Framework Convention on Climate Change, and more than 100 have ratified it—there is a compelling case for more proactive policies to advance the commercialization of solar energy.

With respect to developing countries there are, in my judgment, four elements of a desirable policy: expanded support of solar initiatives by the Global Environment Facility and the international and bilateral institutions providing development finance; the introduction of commercial pricing policies in the developing countries; and a strengthening of both national and international R&D programs.

The Global Environment Facility and Development Finance

A March 1994 international agreement puts the Global Environment Facility (GEF) on a more permanent footing. It is thus now moving from its pilot phase into an operational phase and is well placed to support an expanded program of near-commercial applications of solar energy technologies such as PVs for off-grid and some grid-connected applications, thermal solar projects, advanced biomass power generation projects, and wind power. Some solar projects were financed in the pilot phase. The theory behind such programs is that they will not only test and demonstrate the technologies further and provide operational experience and training, but will also expand applications to help reduce costs. They should also generate backward linkages to the R&D and investment programs of industry and the research community.

An attraction of GEF financing is that it can "leverage" significant finances from other sources, both commercial—private investors, banks, and industry—and public. In its pilot phase, for example, each dollar of GEF resources was supplemented by three dollars from other sources for investments in renewable energy, a "leverage ratio" of 3:1. With declining cost of renewables this ratio is expected to increase to 5:1, and for geothermal and wind energy projects it can go as high as 10:1 or greater (World Bank, 1995).

Commercial Pricing Policies in Developing Countries

There is little doubt that the establishment of commercial pricing policies in developing countries would facilitate solar energy application. For example, the marginal costs of electricity supply are typically 4 cents per kWh at off-peak and 15–20 cents per kWh at peak in urban areas. (Marginal costs at peak are several times above the latter range in some countries because peak demands are spikier.) Average costs are around 10 cents per kWh in urban areas and may range from 20 to 40 cents per kWh in rural areas. Yet average prices often fall far short of these levels (by around 4–5 cents per kWh in 1990), while peak-load pricing is rarely applied, notwithstanding many attempts to introduce it over the past twenty years. Commercial pricing would favor the use of renewable energy in urban and rural areas, for large- and small-scale applications, and would, via peak-load pricing, stimulate the development of short-term storage technologies. In addition, as discussed earlier, it would be a force for energy efficiency.

Strengthened National R&D Programs

Renewable energy still occupies a small share of R&D programs in industrial countries (around 5–7 percent in IEA countries). Although some good programs are in place, a recent study by Keith Kozloff and Roger Dower (1993) has found that the actual levels of R&D are perilously low. It is remarkable that so much has nevertheless been accomplished. But now that the technologies are near-commercial, there is an excellent case for scaling the programs up to include a series of larger-scale demonstration projects. Areas worth investigating include:

—PVs for grid-connected applications, using more advanced materials and design concepts
—advanced thermal-solar schemes (e.g., direct steam generation and higher temperature cycles)
—storage technologies in support of the above
—advanced biomass power generation technologies

Other areas include R&D in transport technologies. R&D needs to be accompanied by tax incentives to encourage producers to experiment with and use the emerging technologies. This is a large subject and requires more analysis than can be provided here. But one point worth making, bearing in mind the circumstances that led to the failure of the Luz scheme to generate electricity through passive solar technology in California in 1991, is the need for a durable policy. It may be recalled that the tax incentives on sales and operating costs for solar energy were continually reduced in unpredictable ways throughout the first six years of the life of the plant. At one point it was proposed to remove them entirely (Lotker, 1991; Ahmed, 1994); the decision was later reversed, after some delay.

The unpredictability and vagaries of public policies ultimately led to the bankruptcy of Luz. The projects have since been taken over by a consortium of U.S. utilities, are being operated under more satisfactory and stable terms, and, from a technical point of view, are functioning very well.

The lesson is that it is better to offer incentives at the time of investment and on capital costs—the grant facility of the GEF works in this way—so that the risks to investors of policies being changed later are reduced. Put another way, investors will heavily discount any incentives provided on recurrent costs and sales, at a rate over and above that usually used to discount the returns to capital and operating risks, in order to allow for the risks of policy reversals. This would not happen, however, if the incentives were provided at the time of investment, and a stronger policy response could be expected from an incentive that in cash-flow terms would otherwise be equivalent to an incentive provided on recurrent costs. If for some reason it were later decided to change policies, then the capital costs of existing schemes would already have been written down, and at least these schemes would be able to continue to operate and the assets would not be wasted. The policy is also consistent with economists' longstanding position of not letting sunk costs lead to inefficiency.

Strengthened International R&D Programs

There is also an economic argument for energy R&D programs to have an international element (see Anderson, 1993b). Between 1945 and 1990, the energy R&D programs of the industrial countries were largely tied to national markets and, in the case of nuclear power, to national security interests. This was appropriate for the political and economic concerns of the period, not least of all because the energy industry was growing rapidly in all countries. But the situation has changed greatly in recent years, and there are two reasons why industrial countries need to strive for more "outward-looking" energy R&D policies.

The first is the global warming problem. If developing countries are to be expected to participate actively in policies to reduce carbon emissions over the long term, it is clear from the scenarios presented earlier that there will eventually need to be a supply of noncarbon technologies with lower costs and continually improved performance becoming available for testing and local demonstration. The GEF and conventional finance will be able to support the application of already tested technologies, but they are likely to make a greater contribution if in addition they have the backing of an R&D program that is capable of putting new and improved technologies into the project pipeline.

The second reason is commercial. Whereas the markets for electricity generation equipment have peaked and declined in OECD countries, those of the developing countries are growing rapidly, more than doubling every decade, and on current trends will add in the next twenty-five years an amount of capacity equal to the entire installed capacity of Europe, North America, and Japan combined. The larger developing economies will be building an industrial base out of such expansions, and both domestic companies and joint ventures with international companies can be anticipated, as is already happening, albeit on a small scale, with the manufacture of photovoltaics. Outlets for manufacturers in the industrial countries are also likely to be sizable. Both commercial logic and environmental concern therefore point to the case for an energy R&D initiative that will establish and support a market for environmentally improved and demonstrated energy technologies in regions where they will be most in demand. To be effective, such an initiative would need to be tied to concrete projects, perhaps mirroring the national programs discussed above.

Conclusions

It has not been my purpose here to propose a specific program so much as to suggest why such a program would be economically and environmentally warranted. Both its technical content and the ways in which it would best be administered deserve further analysis and discussion, as we are at the beginning rather than at the end of a process of defining suitable responses to the global warming problem.

Several institutional arrangements are possible. Some have suggested the consultative group arrangement modeled on the very successful Consultative Group on International Agricultural Research (CGIAR). In addition to supporting local research, CGIAR supports a network of international research centers in Africa, Asia, and Latin America and has had much success in developing new, higher-yielding crop-fertilizer varieties (in implementing the "green revolution") and in its work on crop and livestock diseases. Others suggest a series of bilateral arrangements between the industrial and the developing countries. Another possibility is a publicly supported consortium of manufacturers and energy producers, to develop and administer the types of R&D projects discussed above; the advantage of this arrangement is that it would foster commercial links with, and would draw on the technical and managerial expertise of, the energy industry.

NOTES

1. There are no reliable data on the extent to which the use of oil and gas fuels for cooking has increased.

2. At the international level, see the annual reports of the World Resources Institute. Country studies are also being done. These are being coordinated by the so-called National Environmental Action Plans.

3. See World Bank, 1992: chaps. 6 and 9, for summaries of technologies and costs.

4. The latter involve coal gasification and the use of combined-cycle power plants, in which both gas turbines and steam are used to drive the generators. The improvements in efficiency arise from the higher inlet temperatures to the turbine.

5. However, biomass schemes may be obtrusive and have raised some opposition in Europe.

6. This facility was replenished in March 1994.

REFERENCES

Ahmed, Kulsum. 1994. *Renewable Energy: A Review of the Status and Costs of Selected Technologies.* Energy and Industry Department Discussion Paper no. 240. Washington, D.C.: World Bank.

Anderson, Dennis. 1989. *The Economics of Afforestation: A Case Study in Africa.* Baltimore: Johns Hopkins University Press.

——. 1993a. "Energy Efficiency and the Economics of Pollution Abatement." *Annual Review of Energy and the Environment* 18:291–318.

——. 1993b. "Four Policies and a Proposal for Commercializing Solar Energy Uses in Developing Countries." Paper presented at the World Solar Energy Congress, Budapest, August.

Anderson, Dennis, and Kulsum Ahmed. 1995. *The Case for Solar Energy Investments.* World Bank Technical Series Paper no. 279. Washington, D.C.: World Bank.

Anderson, Dennis, and William Cavendish. 1992. *Efficiency and Substitution in Pollution Prevention: Three Case Studies.* World Bank Discussion Paper no. 180. Washington, D.C.: World Bank.

Anderson, Dennis, and Sean Gammons. 1993. "Energy Taxation and the Environment, with Special Reference to Carbon Taxes and Their Alternatives." Paper presented at a joint IPEC/OPEC workshop, Vienna, September 22–23.

Barnes, Douglas F., Keith Openshaw, Kirk R. Smith, and Robert van der Plas. 1993. "The Design and Diffusion of Improved Cooking Stoves." Unpublished paper owned by author.

Birdsall, Nancy, et al. 1984. *Population Change and Economic Development.* Reprinted from the 1984 *World Development Report on Population.* New York: United Nations.

Bongaarts, John. 1994. "Population Options in the Developing World." *Science* 263 (February 11): 771–75.

Brimblecombe, Peter. 1987. *The Big Smoke.* London and New York: Routledge.

Cochran, Thomas B., Robert Standish Morris, and Kristen L. Suokko. 1993. "Radioactive Contamination of Chelyabinsk-65, Russia." *Annual Review of Energy and the Environment* 18:507–28.

Doolette, John B., and William Magrath. 1990. *Watershed Development in Asia: Strategies and Technologies.* World Bank Technical Paper no. 127. Washington, D.C.: World Bank.

Gregerson, Hans, Sydney Draper, and Dieter Elz. 1989. *People and Trees: The Role of Social Forestry in Sustainable Development.* EDI Seminar Series. Washington, D.C.: World Bank.

Grubb, Michael, Jae Edmunds, Patrick ten Brink, and Michael Morrison. 1993. "The Costs of Limiting Fossil-Fuel CO_2 Emissions: A Survey and Analysis." *Annual Review of Energy and the Environment* 18:397–478.

Heidarian, Jamshid, and Gary Wu. 1994. "Power Sector Statistics for Developing

Countries. (1987–1991)." Draft. Energy and Industry Department, World Bank, Washington, D.C.

Johansson, Thomas B., Henry Kelly, Amulya K. N. Reddy, and Robert H. Williams. 1992. *Renewable Energy.* Washington, D.C.: Island Press.

Kozloff, Keith, and Roger Dower. 1993. *A New Power Base: Renewable Energy Policies for the Nineties and Beyond.* Washington, D.C.: World Resources Institute, December.

Lotker, Michael. 1991. *Barriers to Commercialization of Large Scale Solar Electricity: Lessons Learned from the Luz Experience.* Albuquerque, N.M.: Sandia National Laboratories.

Organization for Economic Cooperation and Development (OECD). 1991. *The State of the Environment.* Paris: OECD.

Organization for Economic Cooperation and Development, International Energy Agency (OECD/IEA). 1993. *World Energy Outlook.* Paris: OECD.

Pearson, Peter. 1988. *Energy Transitions in Less-Developed Countries: Analytical and Practical Understandings.* Paper no. EDP 40. Cambridge, England: Cambridge Energy Research Group, Cavendish Energy Laboratory.

Schipper, Lee. 1993. "Energy Efficiency: Lessons from the Past and Strategies for the Future." Pp. 397–427 in *Proceedings of the World Bank Annual Conference on Development Economics.* Washington, D.C.: World Bank.

Smith, K. R. Kirk. 1988. "Air Pollution: Assessing Total Exposure in Developing Countries." *Environment* 30 (10): 16–20, 29–33.

——. 1993. "Fuel Combustion, Air Pollution, and Health: The Situation in Developing Countries." *Annual Review of Energy and the Environment* 18:529–66.

van der Plas, Robert. 1994a. *A Comparison of Domestic Lamps for Lighting in Developing Countries.* Energy and Industry Series Paper no. 6. Washington, D.C.: World Bank.

——. 1994b. *Solar Energy: Answer to Rural Power in Africa.* Finance and Private Development (FPD) Note no. 6. Washington, D.C.: World Bank, April.

World Bank. 1992. *World Development Report 1992: Development and the Environment.* New York: Oxford University Press.

——. 1995. *Financing the Global Environmental Challenge: A Progress Report on World Bank Global Environment Operations.* Washington, D.C.: World Bank, June–August.

12. Critical Issues in Achieving LDC Environmental and Energy Goals

RICHARD D. MORGENSTERN

In chapter 11 Dennis Anderson attempts to integrate neoclassical economics with technical and policy analysis. He argues that meeting the projected increases in world energy demand, estimated by the World Bank to double or triple over the next three decades, is essential to achieving sustainable development. Such energy growth, he argues, contains the seeds of future lower-polluting activities. He advocates adopting strong market-oriented policies, such as an end to subsidies, as well as developing efficient, low-polluting technologies including a major Global Environment Facility (GEF) solar initiative.

Demand Side

Anderson's policy prescriptions include promoting economic growth, efficient pricing policies, and family planning. Traditional economic rationale supports his logic, but his advocacy of efficient pricing policies has a novel twist: efficient pricing (e.g., ending electricity subsidies) reduces budget stresses, which in turn provides greater funding for health and family planning, thereby leading to reduced population growth. In response to concerns about traditional biofuels, he advocates tree planting, greater use of renewable energy (including photovoltaics in less developed countries, or LDCs), and support for adopting improved wood stoves.

Despite this array of policy prescriptions, however, Anderson is not sanguine about reducing energy demand in developing countries. Citing World Bank studies, he suggests that ending all subsidies, plus imposing a $300-per-ton carbon tax, would reduce energy demand by only about one-fourth. Since the inevitable increase in energy demand leads to more pollution, the only answer is to introduce new technologies.

Supply Side

Anderson contends that decisions encouraging the use of low-polluting technologies have been needlessly belated and that many developing countries seem to be repeating the mistakes of industrial countries. For example, he estimates the total cost of adopting environmentally rational solutions for road transport at about 15 cents a gallon. Costs of electric power particulate-matter (PM) controls amount to approximately 2 percent of total supply costs, and for PM, SO_2, and NOx combined, costs are on the order of 5–10 percent of total costs. He suggests that these costs of new technologies are low when compared, for example, with the estimated $120 billion in annual electricity subsidies in LDCs. Thus, he suggests, ample opportunities exist for "win-win" solutions.

Anderson proposes a "vintage" growth model and notes that the time pattern of substitution of new technologies and capital typically follows an S or logistic curve. Historically, lags have generally been short for PM and lead controls—with exceptions for some countries, such as PM in the United Kingdom—but longer for certain technologies like SO_2 controls and improved cultivation and agro-forestry methods. He expects that solutions to global warming will also experience long lags in diffusion.

Global Warming and a Solar Initiative

The costs and performance of solar technologies have improved dramatically in recent years, with significant future gains expected. The abundance of solar resources and land is well known. Even under current rates of progress, however, it might take two decades or more for solar technologies to mature and diffuse adequately in the developing world. Given the threat of climate change, Anderson suggests that proactive policies are called for. He lays out a four-part program to make the GEF a permanent institution and develop a large-scale solar initiative; to move quickly to market-level pricing of electricity in LDCs; to strengthen national R&D programs through up-front tax incentives; and to internationalize the commercialization of solar development, including creating new partnerships to build an industrial base for solar technologies in developing countries.

Alternative Solutions

Anderson is to be commended for the breadth of the issues he addresses and the comprehensiveness of the solutions he proposes. However, he glosses over two areas where alternative solutions might be considered.

Insurance Policy

In the context of a global decision to reduce greenhouse gas emissions, Anderson proposes a broad-scale "insurance policy" to develop and deploy solar technologies in LDCs. The big question, however, is whether technological development should be driven by market demand for outcomes (in this case low- or no-pollution energy production) or, alternatively, by government (or World Bank) support for particular technologies. Historically, most of the environmental technologies Anderson cites have been driven by market forces rather than by government support of particular technologies (e.g., PM, SO_2, NOx, and lead-free gasoline). While creative use of public-sector resources can speed the development of new technologies, the question is whether the World Bank is appropriately positioned to carry out the mission Anderson advocates.

Joint Implementation

The principal alternative to large-scale World Bank support is to draw on market forces and private capital sources more directly through such strategies as "joint implementation." As envisioned in the Framework Convention on Climate Change, two or more countries could choose to develop a project that would reduce greenhouse gas emissions and then decide how to allocate the reductions among themselves. While there are many implementation issues that remain to be worked out before such arrangements become practical, the expectation is that such projects could be structured so as to tap private capital markets and leave major investment decisions, including the choice of technologies, in the hands of the pubic/private developers of individual projects.

These two approaches are not incompatible. Yet given the likely constraints on publicly supported international lending institutions, the scale of the joint implementation effort is likely to dwarf the World Bank activities.

Conclusions

Anderson does not discuss the potentially enormous gains from properly accounting for environmental damages and resource depletion, or ways of incorporating such thinking into government and private decision making. Recent World Bank studies in Mexico, for example, suggest that certain industries are reducing rather than increasing net economic welfare, as reported in conventional national income statistics. Adoption of such "green GDP" accounting, and its incorporation into national and international lending practices, could lead to enormous changes in national investment behavior.

Moreover, Anderson's focus on reducing emission rates through use of new technologies ignores an equally important source of potential reductions: those derived from consideration and proper pricing of the underlying economic activities that generate emissions. In the case of automobiles, for example, controlling the rate of emissions is an important, cost-effective step. However, one must also consider the growth in vehicle-miles traveled in order to limit long-term vehicle emissions. Marginal cost pricing of fuels, parking, roads, insurance, and related facilities is an essential step in this direction.

Also, while Anderson addresses the electricity pricing issue, he fails to raise the broader set of tax and expenditure questions, including subsidies, embedded in most countries' public finances. Various studies have documented a range of such country subsidies, ranging from energy development and production to broad-based consumer subsidies. Similarly, reform of the tax structure has focused on capital and labor rather than pollution taxes, while likely a bigger issue in developed than in developing countries, may represent an opportunity to make both environmental and economic gains simultaneously.

Finally, Anderson fails to address the indigenous demand for environmental quality. Numerous studies have shown that the demand for environmental quality is an income-superior good. That is, as income rises and so-called basic needs are met, people demand cleaner air, water, and other environmental resources. Governments have an enormous opportunity for anticipating and responding to such "rising environmental expectations" by aggressively educating their populations about environmental issues and developing cost-effective solutions.

Anderson has addressed a broad set of issues relating to economic growth, technological change, and public policy. While I take exception to some aspects of his approach, my concerns are more with omissions from a comprehensive argument that deserves careful study.

13. Perspective and Retrospective on Energy, Environment, and Economy

ROBERT A. BOHM

In 1980 the organizers of the Knoxville World's Fair decided that a Bureau of International Exposition's required educational event, to be sponsored by the fair, should be more than a perfunctory effort. They hoped to contribute significantly to the energy debate that had been raging in the United States and throughout the world since the 1973 OPEC embargo. The result was the International Energy Symposia Series (IESS). As one of the organizers of the three energy symposia that comprised the IESS, I find it interesting to contrast Dennis Anderson's discussion of energy issues (chap. 11) with the way participants in the IESS thought about them in the early 1980s—assuming we thought about them at all.

The IESS was a series of three meetings, each with a specific purpose. Symposium I, the first official 1982 World's Fair event, was held in October 1980. Its task was termed "Issue Identification," a sorting out of the fundamental from the superfluous. Symposium II, held in November 1981, attempted to build on the lessons of Symposium I. Its task was "Issue Analysis," a search for alternative solutions. Finally, Symposium III sought "Issue Resolution," an ambitious task aimed at defining policies that would result in efficient, sustainable future energy development. Symposium III took place in May 1982 while the fair was in progress.

The 1980–1982 Perspective

The IESS produced three volumes of *Proceedings* (Bohm, Clinard, and English, 1981; Clinard, English, and Bohm, 1982; English, Bohm, and Clinard, 1983), which form a permanent record of its deliberations. In order to recapture the flavor of the IESS, I have relied heavily on the contributions of the late Professor David Rose of the Massachusetts Institute of Technology. As Hans Landsberg noted at the time, Rose "managed to capture the spirit and essence" of the meetings.

In setting the stage, we should be reminded of world conditions that formed the milieu of the IESS:

—A barrel of oil cost about $30 in 1980. The power of OPEC to control the price of oil was a very recent memory, although cracks in the cartel were beginning to form. Oil gluts, thought to be temporary at the time, were with us by 1982.

—People with technological solutions to the energy crisis that would be feasible at a $2 nominal price per gallon of regular gasoline commanded attention and had champions. They also spoke of their options as near-term alternatives.

—Three Mile Island was history, but Chernobyl was in the future.

—The Cold War was a fact of life. Mao Zedong was but four years dead in 1980, and Leonid Brezhnev held sway in the Soviet Union. China was considered a mystery, and Russia was an "evil empire."

—Many environmentalists had found a home as advocates of conservation (e.g., rational and effective use) and/or so-called soft energy paths.

—Holiday lights were still scarce in December. At the University of Tennessee, the eternal flame of the Tennessee Volunteer remained dark.

—Americans had given up their love affair with the V-8.

—Then as now, Oak Ridge was a major center for energy-related research and was identified strongly with the nuclear energy alternative.

Battles Won

Holistic Thinking

Dennis Anderson recognizes that energy problems cannot be viewed in isolation. He adopts a global view based on the assumption that energy production and productivity, environmental quality, and economic development are obviously interdependent. He also believes that energy problems have both technological and social aspects that must be incorporated into any solution.

Such holistic thinking was not uncommon back in 1980–82. For example, the following statements are typical of those found in the *Proceedings:* "The availability of energy affects the attainment of many human goals: economic development, environmental quality and, most critically, world peace. . . . In reality [when discussing energy], we are discussing the state of mankind locally, nationally, regionally and globally: who we are, who we wish to be, what is right, what is wrong" (Rose, in English, Bohm, and Clinard, 1983: 10). However, it seemed like we found it necessary to keep reminding ourselves to think holistically. It was far easier to fall into the trap of what Rose (Bohm, Clinard, and English, 1981: 255) called "selective inattention": "I know that global warming is important, and as soon as this fluidized bed system is up and running. . . ."

Hard versus Soft Paths

If Anderson is correct, it is no longer necessary to choose between "hard" and "soft" paths—that is, to decide whether more or less is better, whether centralized energy systems are superior to decentralized ones, or whether increased output of energy (a supply-side solution) is preferable to increased efficiency of energy use (a demand-side solution). Anderson simply declares that high energy demand in the future is a given in both industrialized and less industrialized countries, and that low-energy-demand scenarios are socially indefensible. He believes that the world cannot afford either/or energy solutions. Rather, more of everything will be needed, and specialization should be based on principles of regional efficiency.

Differences Then and Now

National versus International Solutions

In 1980–82 all calls for international cooperation with regard to energy had to be offered as an apology. Today the paranoia surrounding national sovereignty and prerogatives appears to be much reduced. This change is evident in Anderson's discussion of the need for international institution building as contrasted with the IESS discussion of this topic. Energy policy, environmental policy, and economic development policy require an international organization to serve as an honest broker when issues arise that fall outside the purview of a single country. Foremost among the types of organizations needed are those that can serve as mechanisms for cooperation, for example, in the transfer of information, technology, and management skills.

Market-Based Solutions

It is considerably easier to defend market-based solutions and market-based pricing today than in the early 1980s. This change is largely attributable to the end of the Cold War rather than to any increase in sophistication, wisdom, or generosity. For example, in the period immediately following the oil embargo, the budget burden on oil-importing developing countries was viewed as a balance-of-payments problem (i.e., a drain on foreign exchange reserves). As Anderson very novelly points out, however, the problem in fact lay in the failure to pass along the full cost of energy to the consumer. Subsidizing energy prices is the real reason for budget pressure. If markets were allowed to work, the balance-of-payments problems resulting directly from energy imports would disappear. Energy pricing was viewed as both an equity and an efficiency issue in 1980–82. In 1994 the equity aspect of the problem has apparently all but disappeared.

Solar versus Nuclear

At the conclusion of Symposium III, a Final Communiqué was to be issued that synthesized suggested policy solutions from the entire series of meetings. There was little conflict over near-term alternatives and solutions, particularly with regard to the need to use energy resources more efficiently, or over procedural issues. However, in the Draft Communiqué (English, Bohm, and Clinard, 1983: 251–57) a significant role was accorded to solar energy as an ultimate long-term solution. Many symposium participants viewed this solar slant as antinuclear. (Bear in mind that this discussion took place in close proximity to Oak Ridge.) Following long hours of debate, the Final Communiqué (pp. 283–89) emerged with a significantly stronger nuclear flavor but acknowledged the future role of solar energy. The future energy supply option offered—with *future* meaning the twenty- to thirty-year horizon, or about 2000–2010—was "completion of the transition already started, to substantially decreased use of fossil fuels, and to increased use of renewable or virtually inexhaustible energy sources. There are only two major ones: solar power in various forms, and nuclear power in various forms. If the timetable shortens, so much the better" (p. 286).

The communiqué debate was the most acrimonious discussion of the entire symposium series and featured strong statements based on strong feelings. For example: "Solar resources have been utilized for centuries [some for millennia] and provide primary energy for perhaps half the present world population. Today they continue to provide the best energy hope for the future of billions of people" (Draft Communiqué, in English, Bohm, and Clinard, 1983: 256). And, "in saying that the underdeveloped countries must depend upon the sun, we are almost surely committing those underdeveloped counties to a future of unmitigated poverty. . . . I would say that if [solar energy] is the best energy hope for billions of people, then there is little hope for billions of people. And I don't think we ought to concede that point" (Alvin Weinberg, in English, Bohm, and Clinard, 1983: 259–60).

It is interesting that Dennis Anderson does not regard solar energy in the same vein as a cure for the common cold. Rather, he considers solar as a relatively near-term, cost-effective alternative. Furthermore, he does not mention nuclear energy as part of the long-term solution to meeting future energy requirements.

The Trumpet Still Sounds: Some Things Never Change

Despite Anderson's call for market pricing of energy, he still considers the influence of price on behavioral change and tastes to be "secondary." I find this apparent contradiction implausible. Doug Bohi and Joel Darmstadter (chap. 2) corroborate that the 1973–80 years were remarkable with regard to the energy-intensity of consumption. By contrast, 1985–94 has seen considerable backsliding. Anderson's own scenarios attributed a 17 percent drop in demand to achieving efficient energy pricing with price elasticities in the −0.5 range.

Discounting the role of price was a popular pastime throughout the immediate postembargo years. The problem derived from the fact that estimated short-run price elasticities were always quite low in absolute value, as they were a function of in-place capital stock. Although the long-run elasticity might be large in absolute value, its magnitude was unknown because of uncertainty regarding the speed of capital stock adjustment. Changes in energy consumption over the 1973–80 period, however, suggest that the "long run" might well be rather short in duration in the face of substantial relative price shifts. As Bohi and Darmstadter show, ignoring this fact can lead to substantial underestimation of responsiveness to price over a ten- to twenty-year time horizon.

Things Forgotten

Anderson does not discuss equity. Nor does he mention intergenerational redistribution, urban versus rural, or rich countries versus poor countries. In 1980–82 these issues occupied considerable debate time. Rose often referred to the "Just, Participatory and Sustainable Society" (Bohm, Clinard, and English, 1981: 259; Clinard, English, and Bohm, 1982: 15). *Just* referred to fairness, while *participatory* referred to an intragenerational alleviation of inequality.

The word *sustainable* has become very popular in energy and environmental circles in 1994. However, back in 1983 the Final Communiqué of the IESS was entitled "The Outlook for a Sustainable Energy Future" (English, Bohm, and Clinard, 1983: 283). Sustainability is not about keeping the lights on or about preservation for preservation's sake. Rather, it is about intergenerational stewardship of resources. As Robert Solow (1992) has stated, the purpose of sustainability is "to maintain humankind's capital stock, broadly defined to include energy and environmental resources and albeit in often mutating forms, over time intact for future generations so they may improve their standard of living." Defined in this manner, sustainability is about ensuring long-run equity, and about how the so-called three E's— energy, environment, and economy—form the foundation of a stable, peaceful world.

This conclusion should not be forgotten. In the early 1980s considerable emphasis was placed on designing policies to yield full cost pricing of energy that would lead to "rational and efficient use." It was necessary to pursue this approach at that time, but reaching for this goal was not intended to eliminate from consideration the more fundamental question of equity. If equity has taken a back seat for the past several years, it is now time to face it head-on. Failure to deal with inequalities in resource distribution, wealth, or energy in the unstable world of 1994 can have devastating consequences, as the citizens of Bosnia, Somalia, Rwanda, and Haiti can surely attest.

Conclusions

The analysis and discussion contained in the three volumes of the IESS *Proceedings* are as bright, alive, and insightful as if they had been written yesterday rather than over a decade ago. During the debate over that symposium's communiqué, David Rose suggested that what the symposium sought was "the eloquence of John Donne, the cunning of Machiavelli, the wisdom of Lao-tse, all in something that sounds politically acceptable" (English, Bohm, and Clinard, 1983: 277). In retrospect, the participants in those three meetings may have achieved just that.

REFERENCES

Bohm, Robert A., Lillian A. Clinard, and Mary R. English, eds. 1981. *World Energy Production and Productivity: Proceedings of the International Energy Symposium I.* Cambridge, Mass.: Ballinger.

Clinard, Lillian A., Mary R. English, and Robert A. Bohm, eds. 1982. *Improving World Energy Production and Productivity: Proceedings of the International Energy Symposium II.* Cambridge, Mass.: Ballinger.

English, Mary R., Robert A. Bohm, and Lillian A. Clinard, eds. 1983. *Toward an Efficient Energy Future: Proceedings of the International Energy Symposium III.* Cambridge, Mass.: Ballinger.

Solow, Robert. 1992. *An Almost Practical Step toward Sustainability.* Washington, D.C.: Resources for the Future.

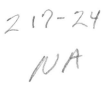

PART FOUR

LOOKING AHEAD
How, and by Whom, Should Future Energy Policy Be Made?

The single most important lesson of the 1970s energy crisis is that there is no technological, economic, or managerial "magic bullet" able to solve our energy problems. Virtually every conceivable option—nuclear power, fossil fuels, biomass, renewables, and energy conservation—has some trade-offs associated with it. These trade-offs are difficult to resolve because every energy option affects a complex array of stakeholders that includes energy producers, consumers, and regulators, each of whom has different aspirations and needs. Moreover, each stands to gain or lose from the different outcomes stemming from these options. This is partly because different options favor different interests. In larger part, however, each of these options raises questions about the appropriate role for governments and private markets.

In order to provide a sustainable, environmentally benign, economical, and abundant energy supply able to support present and future generations, agreement is required not just upon energy sources or their trade-offs but on the relative merit of markets as opposed to government intervention in promoting these sources and reconciling and balancing these trade-offs. As we look toward the future, the challenge we face is forging a political consensus as to the proper balance between governments and markets in promoting a sustainable energy supply while reducing the costs and environmental risks of the components of this supply. If this point has not been made strongly enough thus far, the chapters in this final part drive it home.

The contributors to this part help us to ponder possible answers to this challenge. They generally concur that it is unlikely that another energy crisis can occur, because of the greater flexibility and diversification of global energy supplies. This is a significant change from 1973. They also agree that should another embargo occur, our ability to sustain the shock to energy supplies and economic well-being may be jeopardized by the lack of a sustainable energy system in many industrialized countries. While the Strategic Petroleum Reserve may serve as a "backstop" in case of foreign supply interruption, a benefit to the United States and its allies, by itself it cannot ensure sustainability. Contributors to this part differ over what energy options are sustainable and through which policy path—government intervention or markets—these options should be promoted. This combination of profound disagreement over substantive and strategic means to sustainability mirrors contemporary debates in society at large, and is seen most vividly in the chapters by Worldwatch Institute's Christopher Flavin and the Electric Power Research Institute's Chauncey Starr.

Despite their fundamental disagreement on energy supply options—with Flavin partial to innovative fossil fuel options as a bridge toward renewables and Starr an advocate of nuclear power—both authors favor government intervention to promote the development of these options. Both favor greater federal R&D support and tax or other subsidies. Amplifying the criticisms of government-sponsored energy R&D noted by Denny Ellerman and Glenn Schleede, Starr acknowledges that the nuclear power option suffers from a number of drawbacks. These drawbacks include a historical propensity for choosing wrong or inappropriate technologies to further nuclear energy development, and a tendency to ignore the need for market discipline in establishing the commercial viability of particular reactor designs.

Overall, Starr contends, a sustainable energy supply strategy must consist of technologies that can be rapidly transferred and introduced into countries that lack the money to develop alternative sources of energy indigenously. In the short and medium term, nuclear power remains a viable option, even for these developing nations. In the longer term, however, Starr acknowledges that other technologies must be developed whose impacts are environmentally benign and that are cheaper to operate and maintain.

While Starr acknowledges that the viability of nuclear energy remains limited, in part because of political reservations, one might take his argument further by discussing how widespread acceptance of nuclear energy hinges on the resolution of three issues, each of which almost perfectly exemplifies the wrong-technology-path problem and the lack-of-market-discipline issue: waste management, weapons proliferation, and economical, safe reactor design.

According to nuclear energy proponents, small, modular, passively safe reactors that can be mass-produced and inexpensively assembled may hold the key to restoring public confidence in nuclear energy. Opponents of nuclear power, however, point out that regardless of reactor design, risk-averse utility executives are unlikely to purchase any nuclear technology unless the financial burden of its development can be spread among several parties—an option hindered by federal holding-company rules (Hyman, 1991). In short, there may be no viable markets without considerable government support in such areas as reducing regulatory burdens (e.g., the dual-track construction/operating license path) (Smith, 1991). What makes this issue particularly ironic is that it was government intervention in the nuclear vendor marketplace that initially led to a proliferation of reactor designs back in the 1950s.

Likewise, unless there is assurance that the standards for building, operating, maintaining, and decommissioning these "new" reactors will be enforceable in all countries where these plants are built, public confidence in their safety will be hard to ensure. In short, political intervention in the form of international standards and expectations may be necessary to develop a market (IAEA, 1986).

Attention paid to the issue of nuclear weapons proliferation also reveals this tension. In recent years attention has focused on what might be termed "official" proliferation efforts—efforts to divert fissile materials for use in weapons programs condoned, or deliberately undertaken, by nation-states. This is in stark contrast to "unofficial," nonstate organization actions, such as those of terrorist organizations. Events in Iraq, North Korea, and Pakistan most closely conform to the first problem.

The conventional wisdom is that if these nations can be brought under the existing system of international inspection, monitoring, and safeguards provided by the Nuclear Nonproliferation Treaty—whether through force as in the Iraqi case, or through security guarantees as with North Korea and Pakistan—then the risks of diversion can be brought under control. There are two unresolved problems here that relate directly to the difficulty in reconciling "markets" and "regulation." First, given the decline in superpower control over nuclear markets that has resulted from the end of the Cold War, safeguards are harder to provide because illicit weapons programs are difficult to detect and perhaps even more difficult to deter. This is especially the case in an environment of declining resources (U.S. GAO, 1993). Second, unofficial proliferation threats—or a cross between official and unofficial threats—may be more likely today than in the past. A plausible scenario is for a former Soviet republic, officially or through clandestine groups, to engage in trading fissionable material. Can a government actually "regulate" nuclear materials and a nuclear industry if it lacks the capacity for more basic energy regulation?

The waste issue is the most complex and, in some respects, most intractable of the issues associated with nuclear power—and one that also shows the tension between market-based decision making and command and control regulation. No industrialized nation has yet come up with a politically durable plan for permanent disposal of high-level radioactive wastes, including spent reactor fuel. While the option of deep geologic emplacement remains the most popular method under consideration, several countries, including the United States, are moving ahead with interim plans for aboveground "at-reactor" storage. Unfortunately, public opposition to this option has arisen. Moreover, even if it proves to be acceptable as a short-term (e.g., fifty-year) management option, it is not a viable substitute for permanent disposal of radioactive waste.

While some nations have moved ahead with effective plans for disposal of low- and intermediate-level wastes—or are at least moving ahead with plans for regional cooperation to manage these wastes—this issue too is proving to be contentious. No new sites for the disposal of commercial low-level radioactive waste in the United States or Canada have been opened since the 1970s. Continued public opposition to the siting of new energy facilities, and to the opening of new waste disposal facilities, means that even a new generation of reactors is unlikely to gain support from citizens concerned with the siting of potentially hazardous facilities in "their backyard."

As Barry Rabe convincingly argues in Beyond NIMBY, traditional radioactive waste disposal siting efforts, particularly in Canada and the United States, have relied on a combination of "market" and "regulatory" approaches. The former emphasizes compensating the public for the burdens of waste facilities, while the latter says that governmental authorities should determine the appropriate criteria for siting and then impose an option on a suitable community. Both efforts have failed because they are insensitive to public concerns over risk and fairness and impose a disproportionate burden on select segments of the public:

In the case of market-type approaches, compensation never becomes much of an enticement. It is never seriously explored in many cases, because local animosity toward siting proponents is so great. In the case of regulatory approaches, governmental authorities are rejected as having employed arbitrary criteria to deposit the waste disposal burden of a state, province, or region on a single community. In many cases, . . . target communities are poor and [are composed of] racial minorities, thereby fueling the sense of inequity in distributing these burdens. (Rabe, 1994: 56)

Lest it appear that nuclear power is the only energy source exemplifying these political problems, fossil fuels, an option endorsed by Flavin, also suffer from this dilemma. Flavin wants to see fossil fuels used in more innovative ways. He clearly identifies this tension between markets and government intervention. Some hopeful options include the use of hydrogen, natural gas, and hythane (a combination of methane and hydrogen that could be used to generate electricity and to heat homes). These fuels can be readily introduced into the existing energy transportation and distribution infrastructure and pose little disruption to established markets. They can also be readily adapted by developing nations.

Flavin contends that the greatest challenge in the introduction of such fuels is not technical, but societal and economic. Being able to see a place for these fuels requires more imaginative thinking about energy futures by decision makers than we have previously seen. A natural question is, Why haven't decision makers seen the light? One answer is provided by Flavin's research. No single model can work for all nations. While a competitive market is needed to provide open access to new power sources and to deter monopolistic practices, government incentives are required to induce reliance on diverse, small-scale, and decentralized power sources (Flavin and Lenssen, 1994: 55). However, other questions, whose answers are less clear, also arise.

While it is intuitively satisfying to think that a new energy source can be introduced within the existing transportation and distribution infrastructure, the record of adaptation by this infrastructure on behalf of other energy sources has been far from painless—nor has it been sufficient to induce massive change. For example, electric and methane-powered cars should theoretically benefit from existing access points for fueling these vehicles. Yet, apparently, infrastructure of this type is not enough to increase market viability. Likewise, most successful corporations pride themselves on long-term planning. Nevertheless, if that long-term planning currently includes thinking about alternative energy sources, it is less than clear that this issue is seen as being of grave importance. Perhaps what corporate decision makers need is a "price" signal favoring such long-term thinking. With energy prices currently low, the incentive for such critical thinking is also low.

Kathryn Jackson and Verrill Norwood of the Tennessee Valley Authority (TVA) provide an insightful case study of one energy organization's attempts to meet the challenges of energy sustainability through precisely the type of innovative thinking Flavin endorses. They argue that, in contrast to the more supply-side-oriented market strategies pursued in the past, the federal utility is actively pursuing demand-side management and integrated resource-planning strategies in order to meet the demands of an environmentally challenging business climate for electricity production.

Conceding that TVA is no stranger to controversy, Jackson and Norwood contend that the federal agency has seen previous shifts in its energy policies in response to changing market, political, and technological conditions. In the past, TVA's mandate was building large generation facilities and selling blocks of kilowatt-hours. Demands upon TVA to share its grid with independent power producers, to refine distribution technologies, to improve the reliability of supply, and to promote more efficient energy use are all changing the way TVA and other utilities are doing business. While the outcomes of these demands remain uncertain, we can be assured that the federal utility's future will look very different from its past. Perhaps the most interesting implicit insight of Jackson and Norwood's chapter is that the "signals" that may inspire innovative thinking can have diverse and unexpected sources (e.g., consumer demands, a failed nuclear power program, debt problems).

Finally, the U.S. Department of Energy's Peter Fox-Penner concludes this part by suggesting that we may be entering a new era in which at last there is beginning to be widespread appreciation for the fact that government intervention in energy, as in other policy areas, is a complicated set of choices with many implications. The public demands that government take action to solve problems that cannot be resolved by business alone. However, there can be no viable substitute for market flexibility in making day-to-day decisions about production and use of energy.

By omission, Flavin and Starr raise the same question that Fox-Penner poses: If certain energy technologies are advantageous, then under what conditions and by whom should they be developed? Should government absorb the financial risks of the next generation of passively safe reactors? Or will this put us back into the maw of the "self-promotional" problems that faced nuclear energy in the 1950s? Likewise, if benign fossil-fuel technologies can be developed, should infrastructure be the responsibility of government while fuel is developed by the private sector?

By contrast, promotion of "demand-side management" (as described by

Jackson and Norwood) represents a sort of hybrid policy solution that relies on both economic incentive *and* government intervention (e.g., the Public Utility Regulatory Policies Act, or PURPA, is a federal law that encourages market integration among electricity generators). Here, however, the same need to balance markets and government intervention persists. Unanswered by Jackson and Norwood are such questions as, Should government compel utilities to adopt energy-saving and consumer-demand measures? If the business of a utility is to sell power, is economic incentive sufficient to persuade utilities to urge consumers to conserve? And how do we provide objective, unbiased information to consumers about energy-saving measures, devices, and practices? Should this be left to the marketplace, or should government be the clearinghouse for unbiased information? There is considerable debate on these issues, as testified to by the proliferation of articles on the subject. At the heart of the debate, again, is the question of who should be responsible for policies and programs in demand-side management: government agencies, the marketplace, or others?

Fox-Penner's solution, creatively melding governments and markets, sounds good on paper but is difficult to bring about in practice. This is because, while the public clearly wants some role for government in energy policy—providing a stable, secure energy supply free of foreign threat—and policymakers want to avoid obsessive regulation of energy markets, no policy option has yet drawn together these disparate threads. The very fact that the Clinton administration has, since Fox-Penner's analysis, been forced to "backslide" somewhat on the creative use of policy tools (e.g., a BTU tax) illustrates how the inertia of short-term policy impulse tends to override long-term planning and foresight.

Perhaps the problem with energy policy is that we have yet to decide what's broken so that our "fixes" can make things measurably better. As Fox-Penner notes, the complexity of market behavior and the unanticipated impacts of government intervention suggest the need for "cautious activism" in energy policy. While it remains to be determined what a cautious activist policy would look like, some ideas come to mind (e.g., government "pump-priming" of new energy technologies and support for conservation and energy-efficiency programs; innovative R&D partnerships with the private sector and even other countries).

In conclusion, numerous technologies may serve to put us on an interim path to sustainability. Other technologies and management approaches may warrant a permanent place on our national energy landscape (e.g., conservation, improved end-use efficiency). However, technologies and management approaches are not the problem; politics is. Energy problems and their origins, as the energy crisis of the 1970s shows, present long-term challenges to society. Because policymakers in democratic nations favor shorter decision-making horizons, this challenge is made more difficult by the fact that interest in long-term government/private-sector partnerships must be sustained in the face of shorter-term constraints that constantly intrude on decision makers' freedom to decide.

Perhaps the final lesson in all this is that durable change requires a change in attitude that must, at some level, be shared by decision makers and citizens. To the extent that a literate, educated citizenry accepts the need for a sustainable energy future, their leaders will have the freedom to help shape such a future.

REFERENCES

Flavin, C., and N. Lenssen. 1994. *Powering the Future: Blueprint for a Sustainable Energy Industry.* Worldwatch Paper 119. Washington, D.C.: Worldwatch Institute, June.

Hyman, Leonard S. 1991. "Reviving Nuclear Requires Broad Technology Mix." *Forum for Applied Research and Public Policy* 6 (4): 67–77.

International Atomic Energy Agency (IAEA). 1986. *Final Document Resolutions and Conventions, Adopted by the First Special Session of the General Conference, 24–26 September 1986. GC(SPL.I)/Resolutions.* Vienna: IAEA, October.

Rabe, Barry G. 1994. *Beyond NIMBY: Hazardous Waste Siting in Canada and the United States.* Washington, D.C.: Brookings Institution.

Smith, Sherwood H. 1991. "A Case for Reviving the Nuclear Option." *Forum for Applied Research and Public Policy* 6 (4): 78–82.

U.S. General Accounting Office (U.S. GAO). 1993. *Nuclear Proliferation and Safety: Challenges Facing the International Atomic Energy Agency.* GAO/NSAID/RCED-93-284. Washington, D.C.

14. Myths and Realities in Energy and Environmental Challenges

CHAUNCEY STARR

A reasonable strategy to address global climate change would intensify relevant scientific research; encourage all nations to use economical resource-conserving technologies, particularly for energy use and food; encourage incentives to reduce population growth; support the development and use of all nonfossil energy sources, including nuclear power; and maintain the flexibility to adjust the global energy supply mix as guided by future scientific findings (Starr, 1994).

What can we expect from such a strategy over the next twenty years? With the exception of nuclear power, which many consider heretical, the above objectives are generally achievable. It is likely that only minor shifts in the global energy mix will occur, because it will take decades for capital equipment investment and replacement to be accomplished. Most existing energy facilities last thirty to fifty years and represent significant capital investments that are eventually paid back by their long-term production—which in turn provides the accumulated net earnings available for new capital investments. We should, however, initiate development of advanced commercial technologies now. We can expect a gradual, global spread of a culture of resource efficiency, particularly if economic incentives encourage it. In light of the population and economic growth expected over the next two decades, global environmental challenges will be substantially undiminished. The most promising expectation is that significant strides will be made in our understanding of global environmental dynamics and of the appropriate steps for their mitigation and adaptation. We should be in a position to undertake strategic course corrections in our lifestyles and energy policies with greater confidence.

Environmental Performance

With their larger resource base, today's industrialized nations should be able to perform better than the global average. They can speed the transition to more efficient resource use and environmental remediation. For the past two decades, for example, annual CO_2 emissions from the OECD countries has remained roughly unchanged, while those from Eastern Europe and the former Soviet Union have increased about 2 percent per year and those from the rest of the world about 4 percent per year. The world average increase was about 1 percent per year. Since most future economic growth is expected to come from outside the OECD countries, the future global climate may well depend on the non-OECD world's actions.

While continuing to do the best we can, the United States should recognize that on global environmental issues, our current influence is modest and will decrease over time. While the United States and other industrial countries have historically established extensive energy systems whose replacement is slow, less developed countries offer a relatively virgin opportunity to initiate clean and efficient energy systems, if permitted by their economies, politics, and cultures. Thus, our most effective contribution would be the successful transfer of our advanced technologies to these countries.

The ultimate concern is the effect on individuals' quality of life. As a result, uncertain threats of future climate warming and other global environmental hazards compete for our attention with the pressing realities of today's hazards. In most countries the most urgent environmental hazard is poverty. In all countries, including the advanced industrial nations, economic growth takes precedence in government strategies to provide resources for priorities such as employment, health, education, and security. Environmental issues are part of a lengthy queue of public needs: highly ranked in affluent countries and ranked near the bottom in the poorest, with more urgent needs usually preceding the deferrable. Most would agree on the specific public health and environmental goals that a livable society should seek. But single-issue strategies always come into mutual conflict. Every country faces the challenge of allocating resources and choosing technical options that effectively balance these objectives.

What Should Be Done?

What can we do to face this challenge? The technologist's approach is to choose among feasible alternatives on a comparative benefit/cost/risk basis.

Thus, for energy-related activities, we seek to use fossil fuels more efficiently, with a minimum of emissions and effluents, and to replace these with non-fossil sources wherever practical economics allows us to do so. These nonfossil sources include solar, wind, geothermal, and nuclear. However, a prevailing myth is that solar (including biomass) and wind power can substantially meet global energy growth in the coming decades. These sources should certainly grow from their present minuscule role, but there are many practical reasons why they are likely to contribute only a modest fraction of energy for the foreseeable future (Starr, 1990; 1993). By contrast, the politically unpopular nuclear power option is hardly mentioned in policy discussions in the United States, although it is globally accepted as a significant power source.

We should not try to export our current "politically correct" ideologies. They may be inappropriate for others and only temporary for ourselves. An example is the blatantly antinuclear posture of the environmental movement, presumably based on safety and weapons-proliferation concerns. Such an ideology seriously inhibits the expansion of nuclear power in the United States. The scientific consensus has long been that our nuclear plants are safer for the public than fossil fuel systems. Moreover, they do not emit greenhouse gases. In the United States this antinuclear mythology was bolstered by the Chernobyl accident. The combination of flagrant operator errors and blatant disregard for safety in design that led to Chernobyl no more represents the operation of Western nuclear plants than the Bhopal accident represents the U.S. chemical industry. We haven't shut down our chemical plants.

Regarding proliferation, the hypothetical connection between the nuclear power and nuclear weapons fuel cycles is similar to that between the pharmaceutical industry and chemical and biological weapons. We haven't shut down our pharmaceutical plants. And we have ample evidence (e.g., China, Iraq, Pakistan, North Korea, Israel) that any nation that wants to develop nuclear weapons does not require commercial nuclear plants in order to do so. National security and public safety are obviously important criteria, but ideological opposition merely throttles an energy option that is an inevitable choice in the coming century. Like the rest of the world, the U.S. public will eventually accept nuclear power as another viable energy option.

Conclusions

Social and political support for a pragmatically balanced energy program is hindered by several myths about the future of the environment. These myths have been advanced for a variety of reasons. Because scenarios about the future cannot be proved or disproved, their credibility depends on extending past experiences into the present. The public is presented hypothetical outcomes by media headlines, while only professionals delve into issues of substance. Thus, in the 1970s the Carter administration acted upon the myth of a natural gas shortage and asked Congress to pass the Powerplant and Industrial Fuel Use Act (1978), which denied gas for power plants in order to save it for posterity. The act was repealed in 1987. Clean natural gas is now the most popular choice for generating electricity in the United States.

A serious environmental myth is that economic growth does not require additional energy demand. Between 1973 and 1979 the surge in fossil fuel prices forced energy conservation and temporarily reduced direct use of fossil fuel. However, the worldwide indirect use of fossil fuel for electricity grew in step with economic growth. Since 1979 both direct fuel and electricity use have closely tracked economic growth. Energy use and growth were not separated from one another. Instead, changes in the comparative costs of fuels and energy-efficiency measures caused a shift in the relationship between the two. This is significant because for the foreseeable future, global population and economic growth will result in a substantial increase in energy use, principally in less developed countries, where most population growth will occur.

By the middle of the next century, even with massive conservation and efficiency efforts, global total energy use will be about twice that of today, and electricity production will be about four times that of today (Starr, 1990; World Energy Council, 1993). Even with replacement of fossil fuels by nonfossil sources, as well as active conservation, fossil fuel use will increase.

All this suggests that it is time for energy policy stakeholders to face up to the complex interactions among these issues. Enthusiastic pursuit of a single goal may be psychologically stimulating, but it is not conducive to a balanced national or global strategy. History is replete with the devastating consequences of inflexible adherence to simplistic objectives often voiced by political leaders. But in a domain where technology, economics, sociology, and politics all have equal weight, we need the professional communities to negotiate a balanced strategy for the future.

Where should we be in 2014, and how do we get there? In the short span of twenty years the rate of achieving strategic targets will be determined by public persuasion, rather than by obvious needs. Energy is cheap relative to the costs of investing in long-range environmental programs. Only the least costly investment steps are likely to be taken worldwide, principally in the industrial countries. There are no real strategic pressures on the average individual in any country. Thus, the challenge is convincing people to alter their living patterns and make some sacrifices to reduce present and anticipated environmental impacts. In essence, perceptions of benefit and costs are at the heart of this challenge.

REFERENCES

Starr, Chauncey. 1990. "Global Energy and Electricity Futures: Demand and Supply Alternatives." *Energy Systems and Policy* 14:53–83.
———. 1993. "Global Energy and Electricity Futures." *Energy* 18 (3): 225–37.
———. 1994. *Global Climate, Energy, and Quality-of-Life.* Fifth United States Energy Association Global Climate Change Forum, Washington, D.C.
World Energy Council. 1993. *Energy for Tomorrow's World.* New York: St. Martin's Press.

15. Sustainable Energy for Tomorrow's World

The Case for an Optimistic Future

CHRISTOPHER FLAVIN

Two decades after the oil crisis of the early 1970s, the world is in the throes of rapid change that is sweeping almost every sector. Developments from new information technologies to the rise of genetic engineering and the break-down of traditional social structures are reshaping modern economies and lifestyles at an unprecedented pace. In fields such as computers and medi-cine, few would question that the future will bear little resemblance to the recent past—and probably in ways that we can hardly imagine.

Yet thinking about the world's energy future is surprisingly bereft of the kind of vision found in other areas. Outside of a few pockets of innovation, planners and analysts at leading energy institutions seem trapped in the stagnation and confusion that began with repeated failures in the 1970s and 1980s to develop a viable energy strategy.

Indeed, a review of official energy forecasts and expert studies reveals a rather shocking consensus that during the next few decades only minor changes will occur, yielding a slightly more efficient, marginally cleaner version of today's fossil-fuel-based energy economy. Most governments and international agencies assume that oil and coal will continue to run the world's automobiles, factories, homes, and power plants, supplemented by a little natural gas and nuclear power. According to the experts, any other kind of energy future would be impossibly expensive and impractical (WEC, 1993; OECD/IEA, 1993; U.S. DOE/EIA, 1994a).

The insular world of large energy institutions ignores the forces of out-side change. Throughout the industrial era, energy systems have been pro-foundly shaped by the same kind of technological, political, and economic forces now affecting the world in so many ways. In recent years new tech-nologies for harnessing, moving, and using energy, new public expectations for participation in energy decision making, and a series of environmental crises have become potent forces in the energy marketplace. Together they have the potential to reshape the global energy economy.

Conventional wisdom is sometimes reliable when anticipating smooth trends, but it almost never anticipates major discontinuities. In fact, energy forecasters have had a track record of nearly unblemished failure during the past two decades. Reports by leading institutions in the early 1970s overestimated the amount of nuclear power the world would now be using by a factor of six, while studies in 1980 said oil would cost $100/barrel by the early 1990s (IAEA, 1972–80; Landsberg et al., 1979).

Although corporations and governments are now using more powerful computers and have adjusted their assumptions to account for earlier mistakes, they still seem to be looking at the future through a rearview mirror. The possibility of major technological advances is rarely considered, and change is thought to be an incremental process that only follows well-worn paths. To make matters worse, many assessments are as much a matter of political consensus building as they are serious analysis—telling decision makers what they want to hear, not what they need to know.

The Scenario Approach

In 1968 a team of analysts in the London offices of Royal Dutch Shell were at work on a study of the global oil market. As one of the world's largest petroleum companies, Shell had an obvious interest in anticipating future developments. Like most of the top people at other petroleum firms, Shell's executives comfortably assumed that the future would be much like the recent past—marked by steady or declining prices and growing demand. Yet the Shell planning team, headed by Frenchman Pierre Wack, began to spot signs of trouble in the late 1960s. World oil consumption was skyrocketing, and production in the United States was leveling off. Middle Eastern nations were rapidly increasing their share of the world oil market and beginning to show signs of restiveness about the low oil prices forced on them by Shell and other major oil companies (Schwartz, 1991).

Wack and his colleagues developed two separate projections of future oil trends. The first was a low-oil-price, "business-as-usual" forecast; the second described a future crisis in which OPEC gained ascendancy and oil prices skyrocketed. When his bosses did not react to the new information, Wack abandoned simple numerical projections and developed full-blown "scenarios"—stories about the future that described the reasons for, and ramifications of, a serious oil crisis, including the possibility that major companies' dominance of refining and marketing might be threatened by OPEC. This approach had a far greater impact on Shell executives, who began debating the company's future. Although its strategic plans were not immediately overhauled, when the first oil price hike struck in 1973, Shell was far better prepared than other oil companies. Within a few years it had become the world's most profitable petroleum firm (Schwartz, 1991).

The techniques developed at Royal Dutch Shell have since gained a following among business and government planners, thanks to Peter Schwartz, who worked there briefly in the 1970s and is now president of the Global Business Network, a California firm providing strategic advice to companies. Schwartz points out that scenarios—derived from the theatrical word for script—are not intended to predict the future, a task better left to fortune-tellers, but are "stories about the way the world might turn out tomorrow, stories that can help us recognize and adapt to changing aspects of our present environment . . . [whose] purpose . . . is to change your view of reality—to match it up more closely with reality as it is, and reality as it is going to be" (Schwartz, 1991: 9).

Despite the success of Schwartz and his colleagues at marketing the scenario-planning concept to dozens of companies, its impact in the energy sector has been negligible. Most energy companies and government agencies are still considering the future much the way Shell executives did in 1968. Increasingly powerful computers churn out ever more detailed projections of future trends, but most consider just a narrow band of macro-economic assumptions, often limited to a few different oil price levels and rates of economic growth. They generally assume limited if any technological progress, and major discontinuities or surprises, particularly unwelcome ones, are rarely considered.

The World Energy Council's Scenarios

The World Energy Council (WEC), an organization that includes most of the world's largest private and government energy institutions, typifies the problems. Every three years WEC holds a widely publicized World Energy Congress and prepares an assessment of future energy trends. Its latest effort, published in 1993 after a two-year research effort involving five hundred experts and nine regional teams, foresees an increase in world oil use of some 60 percent by 2020 and a doubling of coal use. The council's baseline scenario, its "most likely," would continue the trend toward heavy use of electricity provided by central power plants. While use of natural gas is expected to increase, so is reliance on the most polluting fossil fuels, including tar sands and oil shale.

WEC envisions the basic outlines of today's energy systems remaining largely unchanged for decades to come. This is an essentially pessimistic view of the world's energy future. The council believes that substantial increases in greenhouse gas concentrations are inevitable and that "the realities suggest that the international energy scene, and many national energy situations will be even more difficult in 2020 than they were in 1990" (WEC, 1993: 154).

The 1993 WEC study considers other scenarios, but they cover a narrow range of assumptions. Even the relatively innovative "ecologically driven" path is tightly constrained. Here, higher levels of efficiency would allow somewhat lower levels of fossil fuel use—and consequent reduction in emissions of carbon dioxide and other pollutants. But the study goes on to question the feasibility of this scenario. Indeed, council representatives have stated that their study proves the impossibility of meeting the central goals of the Framework Convention on Climate Change, which include holding carbon emissions in the industrial countries to 1990 levels and, later, stabilizing atmospheric concentrations of carbon dioxide.

This mischaracterization by WEC of its own results is symptomatic of the problems with most such studies. Although the assumptions used have evolved somewhat since the mid-1970s, they continue to lag behind changes in the real world. Among the blinders that skew all the WEC scenarios are extensive analysis of fossil fuel reserves but no quantitative estimation of renewable resources; lack of a detailed assessment of energy-using appliances and industrial equipment; extensive discussion of the pros and cons of nuclear power, but hardly a mention of solar photovoltaics or fuel cells; and the assumption that developing countries will continue to lag far behind the industrial world in both economic and environmental development.

Overall, these assumptions paint a misleadingly gloomy picture of the potential to forge a more sustainable energy economy in the decades ahead. Sadly, they are the same assumptions used by most governments and by organizations such as the International Energy Agency, the International Institute for Applied Systems Analysis, and the United Nations' Intergovernmental Panel on Climate Change (IPCC).

New Ways of Thinking

New thinking often comes from unexpected quarters. The same rule may apply to technology. Take the telephone, for example, which developed rapidly in the late nineteenth century and then became virtually static in the middle decades of the twentieth. The telephone is again in a period of rapid transition—simultaneously becoming digital, wireless, and portable while also being connected to fax machines, computer modems, and so on.

Energy technologies underwent a period of massive change in the last decade of the nineteenth century and the first decade of the twentieth. The modern energy economy was created virtually out of whole cloth during that period. By 1910 many U.S. cities had been transformed: on the streets, horses had been replaced by automobiles, while candles and gas lights had been supplanted by electric lights. Such transitions are usually driven by an array of social and economic forces, coupled with available technologies that can be applied in new ways.

If these are the conditions needed for rapid change, then the last decade of this century and the first decade of the next may be as revolutionary as those of one hundred years ago. Despite thousands of variables and carefully assembled expert groups, most studies of the world's energy future miss one obvious thing: the energy trends they describe are neither economically nor environmentally sustainable. They will almost certainly be derailed by an array of forces, from political changes sweeping Russia to ecological changes threatening the health of millions. Equally misleading, most energy studies neglect the potential for rapid technological change of the sort now driving advances in electronics, telecommunications, and biotechnology. Few even question the assumption that it will take decades for new energy technologies to be deployed on a significant scale.

The fact is that a thirty-year continuation of the status quo is most unlikely. Indeed, rapid change of many kinds is already taking place, though to see it one has to delve beneath the broad energy statistics that preoccupy most analytical efforts.

Technologies of the Future

One area in which the forces of change have become visible is in the host of emerging energy-efficient technologies. From light bulbs to refrigerators, many of these technologies are at least 75 percent more efficient than the current standard. Even in the power industry, which has sought to improve the efficiency of its equipment for a century, the power plants that opened in the early 1990s are 50 percent more efficient than those of a decade earlier. Such advances suggest that government and international agencies are still overstating future levels of energy use (Flavin and Lenssen, 1994).

The energy landscape has also been altered by the recent emergence of natural gas as the most rapidly growing major energy source. Driven by its environmental advantages and global abundance, gas now seems likely to become the dominant global fuel of the early twenty-first century. This would allow considerable displacement of oil and coal in the near future, and lead to an array of much more efficient and decentralized energy delivery systems. The trend toward gas would continue the move toward cleaner, more versatile fuels that has marked the last century, and counters the conventional view that the world will rely on increasingly dirty fuels in the coming decades.

In addition, recent developments are turning wind power, solar energy, and a host of other renewable resources into economically viable energy options. Some twenty thousand wind turbines are already spread across the mountain passes of California and the northern plains of Europe, while tens of thousands of third world villagers are getting their electricity from solar cells. Solar and wind energy are far more abundant than any fossil energy resource in current use, and declining costs are expected to make them fully competitive in the near future (Flavin and Lenssen, 1994).

The 1990s are marked by another unanticipated technological development: viable alternatives to the gasoline-powered internal combustion engine. Lightweight hybrid-electric vehicles made of synthetic materials and run on gas turbines, fuel cells, and flywheels are about to emerge from engineering labs around the world. With fuel economies three to four times the current average and air pollutants a mere 5 percent of currently allowed levels, these revolutionary new cars, trucks, and buses appear likely to enter the commercial market by decade's end, ushering in an era when automobiles can be refueled at home from the local electric or gas system (Lovins, Barnett, and Lovins, 1993).

The accelerating pace of change in energy systems demands a new paradigm for assessing and managing future developments. What passes for energy analysis today is dominated by preoccupation with resource supplies and the geopolitics of the Persian Gulf, leaving unquestioned the assumptions that we will stay hooked on oil until it is virtually gone and that coal's role must expand simply because it is abundant.

If analysts had held a similarly blinkered view of other sectors, we would still be driving around in buggies and writing on typewriters. The world never ran out of either hay or paper. Rather, people discovered means of accomplishing things more conveniently and economically. The energy sector is no exception. The oil age was ushered in at the turn of the century less by the discovery of petroleum, which occurred much earlier, than by the development of a practical internal combustion engine that made oil useful.

As experts consider the future of the energy economy, they may learn more by studying the electronic revolution of the late twentieth century than by applying the geopolitical and geophysical framework of conventional energy analysis. Most new energy systems will be affected by a range of rapidly evolving technologies, many of which incorporate the latest electronics as a way of raising efficiencies and lowering costs. A variety of stronger, lighter, more versatile synthetic materials will also be applied to everything from wind turbine blades to car frames.

The pace of change will be influenced by the fact that the most important new energy technologies are relatively small devices that can be mass-produced in factories—a stark contrast to the huge oil refineries and power plants that dominate the energy economy today. The economies of mass manufacturing will quickly bring down the cost of the new technologies, and ongoing innovations will be rapidly incorporated into new products, in much the way that today's consumer electronics industry operates.

Sustainable Futures

To consider a "sustainable future" more fully, we at Worldwatch Institute have charted the kind of energy scenario that might emerge if the need for change becomes more urgent. We ask what sort of energy path might unfold if the world decides to take the overriding goal of the Framework Convention on Climate Change seriously: stabilizing atmospheric concentrations of carbon dioxide by the middle of the twenty-first century. The "business-as-usual" scenario used by IPCC shows carbon dioxide concentrations rising from the 1993 level of 357 parts per million to about 700 parts per million at the end of the next century, while WEC has it reaching about 600 parts per million. Our scenario, on the other hand, would have carbon dioxide concentrations leveling off by midcentury at about 450 parts per million, and then declining gradually in the following decades (Pepper et al., 1992; Rotty and Marland, 1993; Marland and Boden, 1993).

This scenario is based on the assumption that governments will rationally but aggressively pursue new technologies. No breakthroughs in cold fusion or government takeovers of the energy sector are considered. Indeed, this "sustainable future" will unfold most rapidly in market-based economies, although governments will need to reverse many of today's energy subsidies. In short, our energy path is characterized by high levels of efficiency, extensive use of decentralized technologies, heavy reliance on natural gas and hydrogen as energy carriers, and a gradual shift toward renewable energy sources (Pepper et al., 1992; Johansson et al., 1993a).

In constructing this picture of a sustainable future, we assume that global economic growth averages 2.5 percent per year, similar to the figure used in most studies. In contrast to them, however, we also assume a slowing of population growth along the lines called for at the 1994 International Conference on Population and Development in Cairo. Higher estimates are inconsistent with the carrying capacity of many regions, particularly Africa, and even at the slower rate assumed here, world population would reach nearly 8 billion before peaking at midcentury. On this path, 85 percent of the population in 2050 will live in what are now called developing countries, up from 78 percent in 1990 (United Nations, 1992; Lewis, 1994).

Virtually all analysts agree that efficient energy use is the cornerstone of a sustainable energy system. In order to hold carbon emissions to about the current level in 2025, and then substantially cut them, we estimate that the world will need to double the current level of global energy productivity— the amount of energy needed to produce a dollar of gross world product— over the next four to five decades. By comparison, it took the United States seventy-two years to double the level of energy productivity it had in 1921 (Schurr and Netschert, 1960; U.S. DOE/EIA, 1994b).

This goal will be made more challenging by the fact that developing countries now have average rates of energy use just one-fifth of Europe's current level and are rapidly building fossil-fuel-powered factories, buildings, and transportation systems. Still, our projected increase in energy productivity is consistent with the potential gains identified in other studies. The technologies needed to accomplish a doubling can already be identified—a goal that seems more possible when it is considered that U.S. labor productivity increased tenfold during the past century (British Petroleum, 1993; Hall, 1994; United Nations, 1993; Maddison, 1987).

A second key to stabilizing carbon dioxide concentrations in the atmosphere is a shift in the mix of fossil fuels. Our sustainable-future scenario posits a 73 percent reduction in coal use by 2025, and a 20 percent cut in oil use, some of which would be replaced by natural gas. Gas is less carbon-intensive than other fossil fuels and lends itself to efficient applications, including potentially widespread co-generation of electricity and heat in factories and buildings. Natural gas resources appear adequate to permit a tripling in global production by 2025. Although such estimates are somewhat speculative, our relatively conservative figures suggest that world gas consumption would peak by about 2030, fall sharply after 2050, and be largely phased out by the end of the twenty-first century.

The third step toward a sustainable future is developing carbon-free energy sources. Although past studies assumed that nuclear fission or fusion would fill this niche—and WEC still makes this assumption—that hypothesis is no longer credible. The current trend in most countries is away from nuclear power. Continuing concern over nuclear waste and safety and the recently uncompetitive cost of nuclear power suggest that this trend will continue. We assume a gradual decline in nuclear power between 2000 and 2025 as aging plants are retired (Lenssen, 1994; Greenpeace International et al., 1992).

As fission fades, nuclear fusion seems increasingly unlikely to ride to the rescue. Scientists working in government-funded fusion programs in Europe and the United States do not foresee it being introduced on a commercial scale before 2030, while skeptics argue that long-term prospects for fusion are dubious. Even if current development efforts succeed, the large power plants that would likely result run counter to the current trend toward decentralized, modular technologies (Crawford, 1990; European Parliament, 1991; Lake, 1992).

Prospects for renewable energy sources are far brighter. Currently, only biomass and hydropower are used in sufficient quantity to show up in world energy statistics, accounting, respectively, for 13 and 6 percent of world primary energy use. These shares are likely to grow in the next few decades in a sustainable future, but in both cases growth will be constrained by resource limitations. At the same time, the so-called new renewables— solar, wind, and geothermal technologies—which currently play a minimal role, are likely to take off within the decade as costs decline and production is scaled up (British Petroleum, 1993; Hall, 1994).

Under our scenario, wind power, solar photovoltaics, and solar thermal would each provide about as much primary energy in 2025 as nuclear power does today. Each would be used primarily for electricity generation, though in different ways. Solar thermal and wind power stations, roughly 1,500 gigawatts each, would be deployed across resource-rich areas such as the U.S. Great Plains, the Thar Desert in India, La Ventosa in Mexico, and the

North African Sahara. Solar photovoltaics would be more widely distributed, mainly on the roofs and facades of buildings.

In order to continue the trend away from fossil fuels after 2025, a 75 percent increase in the harnessing of renewable energy will be needed between 2025 and 2050. By then, renewables would displace oil as the world's second largest energy source, providing over half the world's primary energy, with the share rising as high as 90 percent by 2100.

Although an energy system this different is hard for most people to envision, we see no technical or economic barriers to such a transition. Projected annual growth in new renewable energy technologies—as high as 20–30 percent—is actually slower than the growth rates of nuclear power in the 1960s and 1970s, or of personal computers in the 1980s. By 2025 the renewable energy industry could have annual revenues as high as $200 billion (1993 dollars), or twice the 1993 revenues of Exxon, the world's largest oil company (Spiegelberg, 1992; Young, 1993; *Fortune*, 1994).

A New Energy Infrastructure

How, in liquid petroleum's absence, would energy be stored, transported, and distributed to customers? Most recent efforts to anticipate the energy system of the next century assumed a big increase in the role of electricity and a continuing dependence on liquid fuels for transportation. But electric power is an expensive form of energy that is difficult to store, which is one reason why it meets only about one-eighth of total end-use energy needs today. Relying on electricity as the world's main energy carrier would require using it for many wasteful and inefficient applications such as heating and cooling, which account for over half of current energy use in industrial countries (U.S. DOE/EIA, 1993b, 1994a; Mueller Energy Technology Group, 1991).

Other experts believe that we should try to preserve the current liquid-fuels infrastructure by liquefying coal or biomass and turning them into methanol or another fuel. Yet methanol is a poisonous fuel that readily mixes with water and causes health problems if it comes into contact with skin. Moreover, methanol from coal would exacerbate rather than diminish the greenhouse effect, while methanol from biomass would create other problems. Given the stresses already affecting many biological systems, and the scarcities of cropland and irrigation water that are projected for the near future, heavy reliance on methanol from biomass would likely be disruptive and unreliable. Such scenarios also assume that much methanol would be carried in large quantities from production sites in Africa and Latin America to large energy markets in Europe and East Asia—in effect recreating many of the problems with today's petroleum economy (Fishbein and Henry, 1990; Greenpeace International, 1993; Johansson et al., 1993b; Brown and Kane, 1994).

Another, more appealing vision beckons. Humanity first relied on solid fuels, moving from wood to coal, and then shifted to liquid oil early this century. In recent decades a new trend has developed, with natural gas beginning to displace both liquid and solid fuels in many applications. This shift toward gas is a continuation of a long-term trend toward more efficient, less carbon-intensive fuels—part of the gradual "decarbonization" of the world energy system. Figures for the United States show this broader trend, though recent patterns are partly distorted by the disruption in gas markets that occurred during the 1970s. As natural gas enters the transportation market in the next decade, it will not be excluded from any major segment of the energy economy (U.S. DOE/EIA, 1993a, 1994a).

As natural gas supplies level off or are voluntarily kept in the ground in order to reduce carbon emissions, a substitute will be needed in the decades ahead. The fuel most likely to fill this niche is hydrogen, the simplest of the chemical fuels—essentially a hydrocarbon without the carbon. When it is combined with oxygen to produce heat or electricity, its main emission product is water and some nitrogen oxides.

The logic of a transition to hydrogen has been argued by scientists for more than a century. Although hydrogen has a reputation as a dangerous fuel, this is largely a myth. It can be explosive in the right conditions, as is true of gasoline and natural gas. However, if properly handled, hydrogen is probably safer than the major fuels in use today.

While small amounts of free hydrogen are contained in natural gas, its more common, abundant source is water. Electricity can be used to split water molecules through electrolysis, a century-old technology already used commercially. Although it is relatively expensive today, the cost would come down as the technology was scaled up. The requirements are relatively modest. All current U.S. energy needs could be derived from just 1 percent of today's U.S. water supply. Even in most arid regions, water requirements would not be a major constraint on hydrogen production. The water needed by a photovoltaic power plant producing hydrogen is equal to just 2.7 centimeters of rain annually over an area the size of the plant. And in the long run, hydrogen may be derived from seawater (Ogden and Nitsch, 1993; U.S. DOE/EIA, 1993b, 1994a; Solley et al., 1993).

The real challenge will be finding inexpensive sources of electricity that can be used to split water. Most of the early hydrogen advocates came from the aerospace and nuclear industries. They developed a centralized vision of a hydrogen-powered economy. Some predicted massive offshore nuclear islands that would produce enough hydrogen to serve whole countries. Others suggested that orbiting satellites could beam concentrated solar energy to gargantuan hydrogen production centers on Earth. None of these schemes is remotely practical or economical. All ignore a key advantage of hydrogen: the equipment to produce it is almost as economical on a small scale as on a large one. In production and use, hydrogen lends itself to an efficient, decentralized system (Gregory, 1973).

In our scenario the obvious candidates for hydrogen production are wind and solar energy, supplemented by biomass gasification. As large wind farms and solar ranches appear in sunny and windy reaches of the world, they can be used to generate electricity that is fed into the grid when power demand is high, and be used to produce hydrogen when it is not. Additional hydrogen could be produced in individual homes and commercial buildings using rooftop solar cells. Hydrogen could either be stored in basement tanks for later use or be piped into a local distribution system. In either case, it would gradually fill the niches occupied by oil and natural gas today—including home and water heating, cooking, industrial heat, and transportation. Hydrogen-powered cars have already been developed by several companies, with the main future challenges being an improved storage tank and an inexpensive fuel cell engine (Ogden and Williams, 1989).

The key to practical use of hydrogen is efficiency. According to research carried out at Princeton University and the German Aerospace Agency, even if the cost of the electricity used to produce hydrogen were only 1–2 cents per kilowatt-hour, which may be possible in well-sited facilities built to produce power as well as hydrogen, the delivered cost of the fuel would still be more than three times what U.S. consumers pay for natural gas, though roughly equal to the price European consumers now pay for gasoline (including taxes). But hydrogen lends itself to highly efficient applications, and as a result, the actual delivered costs of energy services to customers would in all likelihood be lower than they are today (Ogden and Nitsch, 1993). One key to efficient use is the fuel cell, which can produce electricity directly from hydrogen at an efficiency as high as 70 percent.

Indeed, the fuel cell may one day be thought of as the silicon chip of the hydrogen economy. Many homes could have reversible fuel cells, capable of producing hydrogen from electricity and vice versa (Abelson, 1990).

The transition to a hydrogen energy system raises other important questions. We currently depend on large quantities of petrochemicals, used in everything from fertilizers to cleaning fluids and clothing. These carbon-based materials are derived from various oil by-products that flow in enormous quantity from refineries. If oil is abandoned, we will need another source. Two options beckon. In the short run, natural gas can be separated into hydrogen and carbon, with the latter going into useful materials rather than the atmosphere. In the long run, we can "grow" petrochemicals on a sustainable basis—a far more important, and less land-intensive, use of cropland than simply having it produce fuels.

Eventually, much of the world's hydrogen is likely to be carried to where it is needed through pipelines similar to those now used to carry natural gas. This is more efficient than the oil or electricity distribution systems in place today. Power lines, for example, are expensive to build, and over long distances they lose significant amounts of electricity. Moving compressed gas is substantially less expensive, though hydrogen will cost 30 percent more than natural gas to move, just because it is a lighter element that is more difficult to compress. In the early stages of the transition to hydrogen, the new energy gas can be added to natural gas pipelines in concentrations up to 15 percent, in a clean-burning mixture known as hythane. Hydrogen could also be produced from natural gas, either in central facilities or right at the gas station. In the long run, engineers believe that it will not be too difficult to modify today's natural gas pipelines so that they will be able to transport hydrogen (Ogden and Nitsch, 1993).

Over time, solar- and wind-derived hydrogen could become the foundation of a new global energy economy. All major population centers are within reach of sunny and wind-rich areas. The Great Plains of North America, for instance, could supply much of Canada and the United States with either electricity or hydrogen. Pipelines that now link the gas fields of Texas and Oklahoma with the Midwest and Northeast could carry hydrogen to these industrial regions. Although renewable energy sources are more abundant in some areas than others, they are far less concentrated than oil, with two-thirds of proven world oil reserves in the Persian Gulf.

For Europe, solar power plants could be built in southern Spain or North Africa. Hydrogen could be transported into Europe along existing pipeline routes. To the east, Kazakhstan and other semiarid Asian republics could supply energy to Russia and Central Europe. For India, the sun-drenched Thar Desert is within easy range of the rest of the country. For the more than 1 billion people of China, hydrogen could be produced in the country's vast central and northwestern regions and shipped to population centers on the coastal plain. And in South America the wind resources of Patagonia could become a valuable energy resource for the entire southern part of the continent.

Many people assume that to produce sufficient hydrogen and other fuels from renewable energy sources requires such huge swaths of land that extensive dependence on them is impractical. In fact, solar and wind energy are far less land-intensive than many energy sources now in use. Today's giant hydro dams and coal strip mines claim extensive land areas that are often rendered unusable for anything else for centuries. Moreover, most of the land needed for harnessing wind and solar energy is in wind-swept plains and deserts, where land values are low (Kozloff and Dower, 1993).

The amount of land required for renewable energy development is surprisingly modest. To provide the world with the 55 exajoules of solar thermal energy needed in 2050, an area about the size of Costa Rica or Bhutan, or less than one-sixth the size of Arizona, is needed. To supply the 50 exajoules of wind energy called for in our scenario, wind farms would spread over an area about the size of Vietnam, or less than the area of Montana. (Only one-tenth of the area covered by a wind farm would actually be occupied by turbine towers and service roads, however, leaving the rest of the land available for crops or livestock.) The land needed for solar and wind development is small enough that environmentally sensitive areas can be withheld from development without significantly diminishing the energy that can be harnessed.

The Challenge for Developing Countries

Developing countries, with more than three-quarters of the world's population, account for just one-fifth of the carbon dioxide released to the atmosphere during the last century, a figure that has risen only to one-third in recent years. Although per capita emissions in industrial countries still average ten times those in poor nations, as the latter industrialize, they become central to any effort to stabilize climate (Marland and Boden, 1993).

In the past, many experts assumed that developing-country energy use (and carbon emissions) would gradually rise toward the levels in industrial countries. But if the world is to hold emissions to the 1990 level in 2025 and then gradually reduce them, another path is needed. Under our scenario, developing-country emissions would increase, but the more profound trend would be a steep decline in emissions in industrial countries, perhaps by as much as half by 2025. The net result is that per capita emissions would converge at one-fifth the current level in industrial countries.

Although many third world energy officials fear that carbon dioxide limits could choke off their economic growth, the greater risk would come in failing to use the more efficient, less polluting energy technologies that industrial nations are beginning to adopt. Pursuing an inefficient oil- and coal-based future would saddle developing countries with a grim combination of uncompetitive technologies and economically draining environmental-cleanup bills. Indeed, Russia has already tried this path to economic development.

Still, there is reason for optimism. When countries industrialize later—as Germany and Japan did in the 1950s and 1960s—they generally leapfrog to a higher, more efficient level of technology. Thus, it is possible that by 2025, countries such as Brazil or Russia could be more energy-efficient than Japan. And it would not be surprising to find that by midcentury, China has a solar-hydrogen system more ambitious than Europe's. The recent opening of many developing countries' energy systems to foreign investment and the elimination of their national monopolies will hasten their catch-up with the richer nations. In the long run, moving away from oil and coal dependence and pursuing a more sustainable energy future is key to the development prospects of third world countries.

Conclusions: The Case for Accelerated Change

How quickly might the transition unfold? When oil prices first soared in the 1970s, energy markets responded slowly at first, then quickened. Government responses were initially misguided, but gradually the more foolish projects were abandoned and better policies emerged. The reaction to a serious climate crisis might well be quicker. The world has been laying the policy and technical groundwork for a new energy system for two decades; with sufficient political pressure, it could accelerate the process dramatically.

Enormous investment opportunities could be created by an energy revolution of this magnitude. Just as the great fortunes of Rockefeller, Ford, and many others flowed from the turn-of-the-century oil boom, the next energy transition promises to create a new generation of successful entrepreneurs. Inevitably, however, there will also be losers. Today's petroleum refineries, oil tankers, nuclear power plants, and automobile factories could soon be seen as stranded investments—many of them representing multibillion-dollar losses once environmental-cleanup costs are accounted for. Major energy companies that fail to anticipate or plan for the coming energy transition are likely to face dim prospects. A recent analogy is the plight of IBM, which failed to anticipate the speed or direction of the personal computer revolution it helped create only a decade earlier—and soon found its blue-chip financial status threatened.

Naturally, the substantial lobbies created by today's oil, coal, nuclear, and automobile industries will initially resist change. Moreover, in most countries, governments are more involved in the energy economy than in any other sector except the military, and their policies are heavily influenced by everything from the job aspirations of German coal miners to the research priorities of French nuclear scientists. The pace of the transformation will be determined in large measure by how quickly the new forces of environmental and industrial change gain the upper hand.

That day may be closer than many experts expect. Despite occasional failures to capitalize on highly touted energy solutions over the past two decades, the technological advances of that period have finally begun to reach a critical mass. History suggests that major energy transitions—from wood to coal to oil—take time to gather momentum. But once economic and political resistance is overcome and the new technologies prove themselves, things can unfold rapidly. If the past is any guide, unexpected events, new scientific developments, and technologies not yet on the drawing board could push the pace of change even faster.

This is how today's energy systems emerged at the end of the last century, and it may be the way a sustainable energy economy begins to emerge at the end of this one. If so, the coming energy revolution will have profound effects on the way all of us work and live, and on the health of the global environment on which we depend.

NOTE

This chapter is based on a Worldwatch book entitled *Power Surge: Guide to the Coming Energy Revolution* (New York: W. W. Norton, 1994).

REFERENCES

Abelson, Philip H. 1990. "Applications of Fuel Cells." *Science,* June 22.

British Petroleum. 1993. *BP Statistical Review of World Energy.* London: British Petroleum.

Brown, Lester R., and Hal Kane. 1994. *Full House: Reassessing the Earth's Population Carrying Capacity.* New York: W. W. Norton.

Crawford, Mark. 1990. "Fusion Panel Drafts a Wish List for the "90s." *Science,* July 13.

Darmstadter, Joel, Perry D. Teitelbaum, and Jaroslav G. Polach. 1971. *Energy in the World Economy.* Baltimore: Johns Hopkins University Press.

European Parliament, Scientific and Technological Options Assessment. 1991. *Study on European Research into Controlled Thermonuclear Fusion.* Luxembourg: European Parliament, July.

Fishbein, Lawrence, and Carol J. Henry. 1990. "Health Effects of Methanol: An Overview." In *Methanol as an Alternative Fuel Choice: An Assessment,* edited by Wilfrid L. Kohl. Washington, D.C.: Johns Hopkins University.

Flavin, Christopher, and Nicholas Lenssen. 1994. *Power Surge: Guide to the Coming Energy Revolution.* New York: W. W. Norton.

Fortune. 1994. "The Fortune 500: The Largest U.S. Industrial Corporations," April 18.

Greenpeace International. 1993. *Towards a Fossil Free Energy Future: The Next Energy Transition.* Boston: Stockholm Environment Institute.

Greenpeace International, WISE-Paris, and Worldwatch Institute. 1992. *The World Nuclear Industry Status Report: 1992.* London: Greenpeace International, WISE-Paris, and Worldwatch Institute.

Gregory, Derek P. 1973. "The Hydrogen Economy." *Scientific American,* January.

Hall, D. O. 1994. Private communication, King's College London, February 18.

International Atomic Energy Agency (IAEA). 1972–80. *Annual Reports.* Vienna: IAEA.

Johansson, Thomas B., et al. 1993a. "Renewable Fuels and Electricity for a Growing World Economy." In *Renewable Energy: Sources for Fuels and Electricity,* edited by Thomas B. Johansson et al. Washington, D.C.: Island Press.

———. 1993b. "A Renewables-Intensive Global Energy Scenario." Appendix to chap. 1 in *Renewable Energy: Sources for Fuels and Electricity,* edited by Thomas B. Johansson et al. Washington, D.C.: Island Press.

Kozloff, Keith Lee, and Roger C. Dower. 1993. *A New Power Base: Renewable Energy Policies for the Nineties and Beyond.* Washington, D.C.: World Resources Institute.

Lake, Gordon J. 1992. "Repatriating Refugees from Reality: Fusion and Fantasy in the European Community." Paper presented at the Joint 4S/EASST 1992 Conference, Gothenburg, Sweden, August 12–15.

Landsberg, Hans H., et al. 1979. *Energy: The Next Twenty Years.* Cambridge, Mass.: Ballinger.

Lenssen, Nicholas. 1994. "Nuclear Power Fades." In *Vital Signs 1994,* edited by Lester Brown, Hal Kane, and David Malin Roodman. New York: W. W. Norton.

Lewis, Paul. 1994. "U.N. Conference to Discuss Plan to Stabilize World Population." *New York Times,* April 3.

Lovins, Amory B., John W. Barnett, and L. Hunter Lovins. 1993. *Supercars: The Coming Light-Vehicle Revolution.* Snowmass, Colo.: Rocky Mountain Institute.

Maddison, Angus. 1987. "Growth and Slowdown in Advanced Capitalist Economies: Techniques of Quantitative Assessment." *Journal of Economic Literature* 25 (2): 649–98.

Marland, G., and T. A. Boden. 1993. "Global, Regional, and National CO_2 Emission Estimates from Fossil Fuel Burning, Cement Production, and Gas Flaring: 1950–1990." Electronic database. Oak Ridge National Laboratory, Oak Ridge, Tenn.

Mueller Energy Technology Group. 1991. *Market Potential for Solar Thermal Energy Supply Systems in the United States Industrial and Commercial Sectors: 1990–2030.* Arlington, Va.: Mueller Energy Technology Group.

Ogden, Joan M., and Joachim Nitsch. 1993. "Solar Hydrogen." In *Renewable Energy: Sources for Fuels and Electricity,* edited by Thomas B. Johansson et al. Washington, D.C.: Island Press.

Ogden, Joan M., and Robert H. Williams. 1989. *Solar Hydrogen: Moving beyond Fossil Fuels.* Washington, D.C.: World Resources Institute.

Organization for Economic Cooperation and Development, International Energy Agency (OECD/IEA). 1993. *World Energy Outlook.* Paris: OECD.

Pepper, William, et al. 1992. "Emission Scenarios for the IPCC: An Update." Paper prepared for the IPCC Working Group I, New York, May.

Rotty, R. M., and G. Marland. 1993. "Production of CO_2 from Fossil-Fuel Burning." Electronic database. Oak Ridge National Laboratory, Oak Ridge, Tenn.

Schurr, Sam H., and Bruce C. Netschert. 1960. *Energy in the American Economy, 1850–1975: An Economic Study of Its History and Prospects.* Baltimore: John Hopkins Press.

Schwartz, Peter. 1991. *The Art of the Long View.* New York: Doubleday.

Solley, Wayne B., et al. 1993. *Estimated Use of Water in the United States in 1990.* United States Geological Survey Circular 1081. Washington, D.C.:U.S. Government Printing Office.

Spiegelberg, R. 1992. Private communication and printout, Division of Nuclear Power, International Atomic Energy Agency, Vienna, March 18.

United Nations. 1992. *Long-Range World Population Projections: Two Centuries of Population Growth, 1950–2150.* New York: United Nations.

———. 1993. *World Population Prospects: The 1992 Revision.* New York: United Nations.

U.S. Department of Energy, Energy Information Administration (U.S. DOE/EIA). 1993a. *Annual Energy Review 1992.* Washington, D.C.

———. 1993b. *State Energy Data Report 1991: Consumption Estimates.* Washington, D.C.

———. 1994a. *Annual Energy Outlook 1994, with Projections to 2010.* Washington, D.C.

———. 1994b. *Monthly Energy Review.* Washington, D.C., March.

World Energy Council (WEC). 1993. *Energy for Tomorrow's World.* New York: St. Martin's Press.

Young, John E. 1993. *Global Network: Computers in a Sustainable Society.* Worldwatch Paper 115. Washington, D.C.: Worldwatch Institute, September.

16. Where Should We Be in 2014 and How Do We Get There?

A Utility Perspective

KATHRYN J. JACKSON AND
VERRILL M. NORWOOD

The Tennessee Valley Authority (TVA) is one of the nation's largest electric power producers, a regional development agency, and a national environmental laboratory. TVA was established by Congress in 1933, primarily to provide flood control, improve navigation, and produce electric power for the Tennessee Valley region.

In 1993 TVA produced 128 billion kilowatt-hours of electricity, more than enough power for three cities the size of New York. TVA's power-generating facilities include eleven coal-fired plants, two nuclear plants, twenty-nine hydroelectric dams, four combustion-turbine installations, and a pumped-storage plant. They provide 25,618 megawatts of dependable generating capacity. In 1993, coal-fired plants produced 76 percent of TVA's electricity, nuclear plants generated approximately 10 percent, and hydroelectric plants were responsible for 14 percent.

TVA provides electric power to 160 local municipal and cooperative power distributors through a network, or grid, of about seventeen thousand miles of transmission lines. These local distributors deliver power to homes, businesses, and industries in the seven-state Tennessee Valley region.[1] TVA also sells power directly to some large industries and federal agencies.

TVA is unique in that it manages natural resources and provides energy. In combining these roles for sixty years, TVA has learned valuable lessons relevant to the global debate on energy and environmentally sustainable development.

How to Get to 2014

Economies are energy-intensive. Energy is at the center of economic development and industrial competitiveness. The challenge for utilities is to produce, deliver, and use energy in environmentally sustainable ways. TVA is striving to meet this challenge, but pressures on the agency make this challenge especially difficult:

—environmental concerns and regulations such as the National Energy Policy Act, the Clean Air Act Amendments of 1990, and the Clean Water Act

—market changes such as co-generation, energy services, and technical services to customers

—competitive pressures such as open access, wheeling (the ability of electric utilities to sell power outside their normal service area), and independent power producers

—increasing demand for greater corporate responsibility

Three major activities designed to meet these pressures are driving TVA into the twenty-first century. These are integrated resource planning, business planning, and environmental management.

Integrated Resource Planning

The purpose of TVA's integrated resource-planning (IRP) process is to develop a clear, comprehensive, long-term energy strategy that meets the Tennessee Valley's electricity needs in a way that reflects consumer values and enhances economic opportunities and the overall quality of life. This process is also designed to preserve flexibility to deal with uncertainty and change; to give TVA and its customers a competitive advantage in a rapidly changing marketplace; and to provide an evaluation cycle of long-term resource requirements focusing on a twenty-year time frame.

To accomplish the IRP process, TVA must fulfill four objectives. The first is to develop a twenty-year energy strategy and short-term action plan that are committed to acquiring needed resources and that provide for monitoring and analysis of results so that modifications can be made quickly and effectively.

The second objective is to develop and implement a state-of-the-art technical IRP process that enables TVA's board of directors to select optimum service options that satisfy customers, that are competitive, and that provide stable rates, environmental stewardship, and economic development. It should also be viewed by national energy opinion leaders as a model. The IRP uses state-of-the-art economic forecasting models to determine future demand for electricity, then compares this information with historical TVA power system data to determine available capacity and to define the gap between capacity and demand. The IRP then evaluates technical options, environmental impacts, scenarios of economic development, and the uncertainties associated with the regulatory environment to generate plans to fill this gap. For TVA to remain a major player in today's highly competitive utility industry, the IRP must generate plans for the future that allow TVA to have the right amount of capacity at the right time and at a competitive cost while minimizing the impact of TVA facilities on the environment.

The third objective is to develop and implement a partnership with TVA's distributors and directly served customers that builds on existing business relationships and enables customers to work with TVA in planning and implementing appropriate components of the IRP to help them meet their own least-cost planning requirements.

Finally, the IRP process must be open so as to afford all interested parties an opportunity to participate. It should generate a range of options, incorporate public opinion in the valley about energy futures and expectations, and build trust and credibility. TVA has established an external review group to bring key stakeholders into TVA's energy planning process. It is composed of eighteen representatives who have been asked to provide their perspectives on the issues confronting TVA as it proceeds through the IRP process. There are representatives of environmental organizations, TVA industrial customers, the distributors of TVA power, and economic development organizations. The members of the review group will be expected to act as conduits to transfer information to the stakeholder groups they represent. It will also ensure that the concerns and needs of ultimate end-users of TVA power are heard, and it will build constituencies for TVA and its distributor customers, provide information about TVA's power system, and comply with the National Energy Policy Act.

Business Planning

The 1990s will see profound changes in the U.S. electric utility business, which has long focused on building larger generation and transmission systems and selling kilowatt-hours. Customers are the new focus, and energy service, not kilowatt-hours, is the product. The National Energy Policy Act, the Clean Air Act Amendments of 1990, and the Clean Water Act all require this sort of change.

TVA recognizes that various technology-sector advancements will provide the means to accomplish this change. The challenge is to expedite technologies and to align TVA plans with future market and political realities. TVA is focusing on integrating its capabilities more fully with the Tennessee Valley region and the nation.

In the area of customer service, TVA offers a wealth of technical expertise to industries and communities in the valley. Some are interested in using co-generation technology to meet their growing need for a cost-effective electricity source. Others want to improve efficiency or reduce wastes.

Customer expectations and end-use technologies are changing, making it increasingly important to enhance the value of electricity services, not just to supply kilowatt-hours of electrical energy. In particular, the growing use of microprocessor-based equipment has created a new demand by utility customers for the highest-quality power available.

TVA is working with its end-use consumers, directly and through its power distributor network, to identify new electrotechnologies that improve overall production or efficiency. Some electrotechnologies increase the intensity of electricity use through wider application of electrical processes. Others decrease it through productivity gains. Induction-heating processes are an example of an electrotechnology that improves production and addresses environmental problems. Microwave disinfection of hazardous medical wastes is another electrotechnology that can help solve a customer's environmental problem.

The more competitive utility industry now emerging has at its core a direct customer focus. Characteristics being encouraged and legislated at the national and state level include wholesale competitive markets for generation, an open-access transmission grid, wholesale and retail delivery competition, and profitability based on service. These are accelerating the innovative application of both new and existing technologies, all intended to provide a competitive edge as viewed from the perspective of customer satisfaction (and in many cases customer competitiveness).

Independent power producers (IPPs) are a major driving force behind the shift to a more competitive environment within the U.S. electric utility industry. A challenge for all utilities is to take advantage of the flexibility and innovations provided by IPPs, co-generators, and nonutility generators while providing high-quality, reliable service for customers. One advantage IPPs have over conventional utility operations is that each IPP plant is a business unit where plant managers have substantial authority to run operations and make a profit.

Many utilities, including TVA, are currently taking aggressive action to cut costs and improve competitive position. The influence of IPPs on the electric utility industry will increase over the next twenty years.

Corporate Citizenship and Environmental Management

Conservation and the thoughtful use of natural resources is a high priority for utilities and a way of life for TVA. A variety of new technologies, such as renewable energy resources and electric vehicles, can revolutionize the way TVA manages the resources in the valley and thus place TVA at the forefront of national resource management efforts.

Renewables such as biomass residue and energy crops may significantly reduce TVA's emissions of greenhouse gases and possibly improve land use and water quality. Demand-side management can help reduce energy consumption and provide opportunities to reduce daily peak energy demands.

Environmental health, welfare, and safety concerns provide both opportunities and challenges for utilities. TVA's emerging technology-based strategic plan takes these concerns into account—from cleaner power plants to more efficient end-use appliances and industrial electrotechnologies. The technology strategy for TVA will define a path to improve the performance of existing facilities, develop new capabilities to generate and distribute electricity in the most efficient manner, serve as the basis for managing demand and energy conservation, and develop new electrotechnologies that are more productive and effective than current industry practices. In this manner TVA can satisfy the valley's energy demands while ensuring a brighter future for our business.

Environmental technology development, which historically emphasized relatively inefficient, end-of-pipe controls, is evolving toward systematic prevention or elimination of pollution. One significant change is the growing realization by utilities that environmental and related concerns are business opportunities, not threats. New electrotechnologies to take advantage of these opportunities, from electric vehicles to the use of coal combustion by-products, are receiving high priority. Several programs are under way within TVA to respond to the clean-environment business criteria and to realize our environmental leadership goal. TVA's coal combustion by-product utilization program is one example.

Each year TVA's fossil plants generate about 3.5 million tons of ash and another 1 million tons of scrubber gypsum. The management of these tremendous quantities of by-products presents a challenge and an opportunity. TVA is challenged to find new, environmentally sound uses for these materials and to fulfill an ambitious goal of utilizing 100 percent of the by-products generated at our plants by the year 2020.

Autoclaved cellular concrete, which uses up to 70 percent fly ash in a concrete mix, offers a promising new way to increase substantially the use of some of the fly ash produced each year. We hope that this technology will attract investors who are willing to go into commercial production at a site in the Tennessee Valley—starting a new industry that can provide more jobs for Tennessee Valley residents.

Finally, TVA operates a national laboratory for environmental research: the Environmental Research Center. This facility has a central mission in TVA's strategic plan. One large program currently being conducted by the center is the air-quality research program, which will provide a scientific basis for regulations and environmental technology in coming decades. This research focuses on developing a sound understanding of ozone exposure, the relationship between exposure and impacts, source contributions to exposure, and the likely benefits to the Great Smoky Mountains National Park (GSMNP) from a reduction in nitrogen oxide (NOx) emissions at TVA facilities. Understanding developed from the GSMNP study is expected to be applicable to other Class I areas in the southern Appalachians and to the southeastern United States.

Conclusions

Even optimistic conservation scenarios conclude that for at least the next half-century the United States will continue to depend on a fossil-fuel-based energy economy. The limited availability of oil and gas is expected to ensure that coal, which represents over 80 percent of the world's fossil energy resource, remains the predominant fuel during this period, with the United States, Russia, and China holding the largest reserves. The issue, then, is not whether coal will be increasingly used but rather how efficient the technology for its use can be, how rapidly the energy economy can increase its reliance on alternative sources, and how conservation and technologies to improve efficiency can reduce growth in energy demand.

During this next twenty years we should try to move beyond fossil fuels to a new energy foundation that depends on sustainable energy and its efficient use. Utilities are up to the challenge. We at TVA look forward with anticipation to the challenge and opportunities of the next twenty years.

NOTE

1. The seven states are Alabama, Georgia, Kentucky, Mississippi, North Carolina, Tennessee, and Virginia.

17. Instruments and Tools in Clinton-Era Policies

PETER S. FOX-PENNER

It is wrongly assumed that government interest in energy policy began in response to the 1973 OPEC embargo. While the facts suggest otherwise, it is still appropriate to use the embargo's twentieth anniversary as an occasion to reflect on the past and to consider the future of U.S. energy policy.

Many retrospectives attempt to divine the future course of policies. For example, a prescient 1982 Congressional Budget Office study reported that historically, the energy problem has evolved through two eras. In the first, no distinct problem was recognized, and energy policy became the de facto outcome of other concerns. In the second, the energy problem emerged as an oil security problem, and energy became a distinct policy arena. Much evidence suggests that the nation is now entering the third era of the energy problem: that of concern for economic efficiency.

Other retrospectives examine scenarios or urge the adoption of new policies (e.g., Ford Foundation, Energy Policy Project, 1974). I abstain from specific predictions, scenarios, or policies and focus instead on the relationship between the *objectives* of federal policy and the *instruments and tools* used to implement it. These terms come from macroeconomics, a field in which the relationship between tools and objectives has long been debated. I draw an analogy between the evolution of the tools-objectives relationship in macroeconomics and in energy policy. In so doing, I argue that we have reached an especially interesting stage in the evolution of energy policy tools, which makes predictions particularly difficult.

Tools and Objectives in Macroeconomics and Energy Policy

For more than a century there has been widespread agreement that the main objectives of macroeconomic policy are to maintain economic growth and high levels of employment and to prevent large bouts of inflation or price instability. Substantial disagreements begin to form at the next "level" around such fundamental questions as how large a role government should play toward these ends, and what specific policies and mechanisms are appropriate.

Responses to these two questions tend to correlate. Policymakers who favor a stronger government role tend to favor more aggressive and interventionist instruments; those who favor a weaker government role tend to favor less intervention. The tide has swung between those who advocate a minimal government role and stable monetary policies and those who call for an active fiscal policy, redistributive taxation, and less emphasis on monetary policy.

A brief chronology might begin in the laissez-faire 1920s, which was followed by Keynesian activism and the New Deal during the 1930s and 1940s. The 1950s saw a renewal of faith in free markets and stable, interest-pegged monetary policies. In the 1960s we returned to fiscal activism and the Great Society. When inflation and stagflation entered the picture, we returned to stable monetary policies—now measured directly rather than via interest rates—and the New Classical School emerged as the antithesis to Keynesian economics. The New Classical School of macroeconomists believe that, in general, macroeconomic changes reflect an aggregation of individual (microeconomic) industry and consumer actions. This school builds models that incorporate "rational expectations" and other rational economic optimizing behavior (Stein, 1994). At the end of this chaotic journey, we find ourselves in an era that has been called one of "cautious activism without simple rules" (Belongia and Garfinkel, 1992).

Current U.S. energy policy can be viewed similarly. Most policymakers agree that the widespread availability of energy at reasonable, moderately stable prices is essential. Since the advent of environmental concerns, most would also add that energy use must reflect a strong preference for environmental quality.

While these objectives are widely shared, disagreements concerning the proper role of government in the energy sector have sometimes seesawed violently from era to era. These disagreements have been fueled by debate over the character of energy—that is, the extent to which it is a public or private good—and, if partly public, the nature of its impacts and the appropriate level and type of government intervention.

This disagreement is reflected in the degree to which we have allowed mostly free markets to distribute energy products. To the extent that energy was "too important" to leave to the marketplace, or energy forms had characteristics that required regulation, markets were not permitted to function alone. To cite an early example, Daniel Yergin writes of Winston Churchill's firm belief in the quasi-public character of energy, in this case oil:

> On July 17, 1913, Churchill, in a statement to Parliament that the *Times of London* described as an authoritative presentation on the national interest in oil, took the idea one step further. "If we cannot get oil," he warned, "we cannot get corn, we cannot get cotton and we cannot get a thousand and one commodities necessary for the preservation of the economic energies of Great Britain." In order to assure dependable supplies at reasonable prices—because the "open market is becoming an open mockery"—the Admiralty should become "the owners or, at any rate, the controllers at the source" of a substantial part of the oil it required. It would begin by building up reserves, then develop the ability to deal in the market. The Admiralty should also be able "to retort, refine . . . or shrink from making this further extension of the vast and various business of the Admiralty." Churchill added, "On no one quality, on no one process, on no one country, on no one route and on no one field must we be dependent. Safety and certainty in oil lie in variety and variety alone. (Yergin, 1991: 160).

About the same time Churchill wrote this, gas and electric utility regulation became commonplace in the United States—an alternative to government ownership, which had become the norm in most of the rest of the world.[1]

Following the war, the Eisenhower administration sought greater reliance on the marketplace (Goodwin, 1981). The administration favored deregulation or state regulation of natural gas and voluntary oil import guidelines, and it encouraged a private rather than government-owned nuclear power industry.

The 1960s and 1970s saw a gradual shift toward greater government involvement, culminating in the National Energy Conservation Policy Act of 1978. This act included new taxes, tax credits, public utility regulatory policies, federal restrictions on fuel use, and other interventionist tools such as mandatory energy-efficiency standards for home appliances and extension of natural gas price controls.

During the 1980s energy policies shifted dramatically toward reduced intervention. Many provisions of the National Energy Conservation Policy Act expired or were repealed, natural gas production was deregulated, wholesale electric markets became more competitive, and political concern over oil imports and many other energy issues reached a low point in public opinion polls. As a complement to reduced interest in energy policy, strong regulatory tools fell into disfavor. The political climate made taxation and heavy-handed government policies particularly unpopular, and a growing constituency argued that "market-based" tools such as emissions trading were far preferable to the instruments of "command and control." These ideas surfaced as early as 1977 but were given a boost by the *Project '88* report produced at Harvard's John F. Kennedy School of Government by former senators Tim Wirth and John Heinz (Wirth, Heinz, et al., 1988). As many "traditional" concerns ebbed, the issue of environmental preservation steadily rose in importance. The link between energy use and environmental quality created a new justification for federal policies, though the proper form of response is still subject to much debate. Meanwhile, U.S. oil imports gradually returned to the levels of the 1970s, and the domestic gas and oil industries suffered greatly from weak economic growth and low prices. The decade concluded with the emergence of twin global environmental concerns: the "ozone hole" and climate change.

Developments in the 1980s, particularly the joining of energy and environmental policy, set the stage for a return to activism in the 1990s. The new era began before the advent of the Clinton administration, when Congress amended the Clean Air Act and enacted the Energy Policy Act of 1992. The Energy Policy Act was by far the most sweeping energy legislation since 1978, but it contained only a small amount of the strong medicine found in its predecessor. It created a few new tax incentives and credits and modestly increased energy codes and standards for equipment and buildings. However, it also encouraged market reliance. In particular, it attempted to open wholesale electric markets.

Observations on the Present

Today we find ourselves defining a light-handed, market-oriented federal role in energy markets. While I am hardly an impartial observer, I believe that the Clinton administration can be characterized as fairly activist. However, it is advancing an ambitious energy and environmental policy agenda with less intervention and more market tools than previous activist administrations.

A sure sign of its activism is the administration's belief that the federal government's role in energy policy is greater than precommercial R&D of new energy products. Although the U.S. Department of Energy (DOE) and the administration have a strong, expanding commitment to R&D, there is also a mandate to move technologies "out of the labs and into the market-place." This implies a belief that markets alone will not adequately meet the administration's goals of creating jobs, improving the balance of trade, and providing energy security and environmental quality. The greatest single example of this approach is found in the president's Climate Change Action Plan (Clinton and Gore, 1993). This plan calls for a reduction of greenhouse gas emissions to 1990 levels by the year 2000—a firm, ambitious goal to halt a century-long trend. Of the forty-three measures proposed in the plan, only one required new legislative authority, and it did not create a new "command and control" role.[2] The remaining measures in the plan relied either on existing provisions of the Energy Policy Act or on voluntary energy-efficiency and pollution prevention programs in the self-interest of participants.

Some of the plan's measures are experiments in market-sensitive industrial policy in the energy sector. In the 1970s the federal government's approach to encouraging solar energy consisted of tax credits and the creation of large-scale solar energy facilities. The Climate Change Action Plan takes a different approach. The federal government will contribute relatively modest sums ($18 million in fiscal year 1995) to a "buyers' club" of electric utilities that agree to buy and install small blocks of solar-electric capacity on their systems. The "volume buy" is intended to induce the solar manufacturing industry to make capital commitments sufficient to produce in volume at lower cost. The federal contribution is designed to pay only for the wedge between present prices for solar-electric cells and the price of alternative power sources.

This approach is a potentially elegant way of pulling a crucial industry along the learning curve and reducing new-technology risk. The approach's success hinges on funding the correct wedge over the correct period of time without creating a subsidy-dependent industry. In addition, the buyers' club will succeed only if several manufacturers make capital commitments so that no one winner dominates the industry over time.

Another example of the new approach is the contrast between the interstate highway system of the late 1950s and modern-day programs to create an infrastructure for alternatively fueled vehicles (AFVs). In the first instance we embarked on a massive federally funded program financed by its own tax and trust fund. Today we operate a far less costly program that attempts to induce fleet operators in major cities to convert to AFVs voluntarily. Federal government vehicles are converted to AFVs and placed in cities where other vehicle operators will match them, thus encouraging fuel suppliers to install fueling facilities.

Implications of the New Approach

These two examples illustrate a policy approach that might, like its macro-economic cousin, be termed "cautious activism without simple rules." The cautious-activism (CA) approach symbolizes an era in which economic thinking has permeated energy and environmental policy. And while it is obviously too early to declare a permanent shift, several implications of the new approach are becoming apparent.

One implication of CA is the need to study carefully the operation of energy markets and the institutions and players that affect it. Energy service markets have become increasingly complex, and many market participants play a role. Effective federal intervention requires that we analyze all the activities and players and then move to fashion an involvement that addresses a need or barrier that markets alone cannot address. Moreover, involvement must be uniquely federal and must complement the work of others struggling toward similar goals. As an example, gas and electric utilities and many state agencies operate energy-efficiency programs. Currently, the collective efforts of these groups vastly outstrip federal funding and effort for the activities.[3] The federal intervention we envision in this area consists of encouraging expansion of the coverage and comprehensiveness of these programs, helping to disseminate "best practices," and recognizing and rewarding our public- and private-sector partners.

A second implication of this approach is the need to develop tools that measure the impact of our activities. Elegant alliances between the federal government, private firms, and other entities may be good policy, but if there are no ribbons to cut, federal jobs to count, or other tangible signs of success, how will we demonstrate the value of our efforts?

DOE is spending considerable time looking into this question. We are producing measurement tools that quantify the impacts of various federal programs, drawing on the work of public- and private-sector researchers. It is too early to report results, but we believe that substantial improvements can be made by developing high-quality data and using carefully developed techniques and assumptions to analyze it.

Conclusions

As with any combination of policies and tools, the future of CA energy policies rests on their perceived effectiveness and on what Davis (1978) calls the "general political milieu." Whatever the outcome, however, we are learning interesting lessons about the implications of this approach for formulating and executing policies. On the plus side, these sophisticated policies require more compelling justification and a higher standard of care. In the minus column, CA policies can be difficult to communicate and may call for more and better resources than governments can muster.

Perhaps most important, policies that interact with the market must be fast and flexible. Accommodating this concern calls for stronger and more flexible relations with our stakeholders, including Congress. We need a generation of federal policymakers who eschew the not-invented-here syndrome and look to the marketplace and other participants for reality checks and inspiration. Our understanding of technology must be sound, but so must our understanding of markets, the financial community, and the mechanisms of public choice. We must avoid allowing what is light-handed today from becoming tomorrow's unchanging entitlement. In our methods, our targets, our relations with Congress, and our self-justifications, we must—to close by quoting President Bill Clinton—learn to make change our friend.

NOTES

I wish to thank Susan F. Tierney, Phil Meguire, David Levine, Indermit Gil, George Hall, David Kovner, Joe Easton, and Lorraine Cox. All errors are mine alone.

1. Kahn's (1989: 3–6) list of factors that influenced successive jurists as to the need to regulate utilities bears a surprising resemblance to the "instruments and targets" questions that often occupy macroeconomic policymakers.

2. The measure requiring legislative action is a change in the federal tax code to discourage employer-paid parking. The plan also calls for altering federal licenses for hydroelectric sites in order to increase production. The latter should not require new legislative authority. See Climate Change Action Plan: Technical Supplement (U.S. DOE/OPPPE, 1994: 50).

3. Electric utilities spent about $2.2 billion on energy-efficiency programs in 1992, and gas utilities also spent considerable sums. State energy offices spent an additional $10–50 million. In contrast, the fiscal-year 1994 energy-efficiency budget of DOE's Office of Energy Efficiency and Renewable Energy is on the order of $400 million, including the low-income weatherization program.

REFERENCES

Belongia, Michael, and Michelle R. Garfinkel. 1992. "What Is a Business Cycle?" In The Business Cycle: Theories and Evidence, edited by Michael Belongia and Michelle R. Garfinkel. Boston: Kluwer Academic.

Clinton, William J., and Albert Gore, Jr. 1993. The Climate Change Action Plan. Washington, D.C.: The White House.

Courrier, Kathleen. 1980. *Life after '80: Environmental Choices We Can Live With.* Andover, Mass.: Brick House Publishing.

Davis, David Howard. 1978. *Energy Politics.* 2d ed. New York: St. Martin's Press.

Duchesneau, Thomas D. 1975. *Competition in the U.S. Energy Industry.* Cambridge, Mass.: Ballinger.

Dunkerly, Joy. 1980. *International Energy Strategies: Proceedings of the 1979 IAEE/ RFF Conference.* Cambridge, Mass.: Oelgeschlager, Gunn, and Hain.

Ehrlich, Everett M., and Raymond C. Scheppach. 1982. *Energy-Policy Analysis and Congressional Action.* Lexington, Mass.: D. C. Heath.

Executive Office of the President, Council on Environmental Quality. 1990. *Environmental Quality: Twenty-first Annual Report.* Washington, D.C.: U.S. Government Printing Office.

Ford Foundation, Energy Policy Project. 1974. *A Time to Choose: America's Energy Future.* Cambridge, Mass.: Ballinger.

Goodwin, Crawford D. 1981. *Energy Policy in Perspective: Today's Problems, Yesterday's Solutions.* Washington, D.C.: Brookings Institution.

Kahn, Alfred. 1989. *The Economics of Regulation.* Cambridge: MIT Press.

National Academy of Sciences, National Research Council. 1980. *Energy in Transition, 1985–2010.* San Francisco, Calif.: W. H. Freeman.

Peltzman, Sam. 1975. "An Evaluation of Consumer Protection Legislation: The 1962 Drug Amendments." *Journal of Political Economy* 83 (3): 663–67.

Silk, Leonard S. 1975. *Contemporary Economics: Principles and Issues.* 2d ed. New York: McGraw-Hill.

Stein, Jerome L. 1994. "Monetarist, Keynesian, and New Classical Economics." *American Economic Review* 71 (2): 139–44.

U.S. Department of Energy, Office of Energy Efficiency and Renewable Energy (U.S. DOE/OEERE). 1994. *Implementation of the President's Climate Change Action Plan.* Washington, D.C., March 23.

U.S. Department of Energy, Office of Policy, Planning, and Program Evaluation (U.S. DOE/OPPPE). 1994. *The Climate Change Action Plan: Technical Supplement.* DOE/PO-0011. Washington, D.C.

Wirth, Tim; Henry J. Heinz III; et al. 1988. *Project '88: Harnessing Market Forces to Protect Our Environment—Initiatives for the New President.* Cambridge: John F. Kennedy School of Government, Harvard University, October.

Yergin, Daniel. 1991. *The Prize: The Epic Quest For Oil, Money, and Power.* New York: Simon and Schuster.

Appendix A. Proposals and Actions Listed in Presidential Energy Messages and Statements, Nixon and Ford Administrations

Presidential Messages and Statements Summarized in Table (see p. 141)

(see p. 141)

President Richard Nixon

June 4, 1971: President's energy message

February 15, 1973: President's 1973 State of the Union message on natural resources and environment

April 18, 1973: President's energy message

June 29, 1973: Statement by the president on energy

August 9, 1973: Proposed Mandatory Oil Allocation Program (Energy Policy Office)

September 8, 1973: Remarks by the president on the nation's energy policy

October 2, 1973: Mandatory Propane Allocation Program (Energy Policy Office)

October 9, 1973: Statement by the president on the need for conservation of heating oil

October 11, 1973: Statement by the president on energy R&D

October 12, 1973: Mandatory Middle Distillate Allocation Program (Energy Policy Office)

November 8, 1973: President's energy emergency address

November 25, 1973: President's radio and television address on national energy policy

December 4, 1973: Statement by the president on the Federal Energy Organization

December 19, 1973: Announcement by the president of a proposed Windfall Profits Tax

January 15, 1974: Mandatory Petroleum Allocation Program (Federal Energy Office)

January 23, 1974: President's energy message

PROPOSALS AND ACTIONS LISTED IN PRESIDENTIAL ENERGY MESSAGES AND STATEMENTS - Nixon and Ford Administrations

Proposal or Action	71 6/4	2/25	4/18	6/29	'73 8/9	9/8	10/2	10/9	10/11	10/12	11/8	11/25	12/4	12/19	'74 1/15	1/23	3/6	4/19	6/14	6/25	7/2	8/8	10/11	10/15	'75 1/15	1/15	5/27	7/14	'76 2/26	2/22	10/28
Research, Development & Demonstration																															
$10 billion-5 yr. energy R&D program	X																														
Recommend energy R&D program (Chairman of AEC)			X	X											X	X								X					X		
Liquid Metal Fast Breeder (LMFBR) demo	X		X	X											X	X													X		
Nuclear reactor safety & supporting technology	X		X	X											X	X													X		
Advanced reactor concepts (gas cooled, molten salt, light water)	X														X																
Gas cooled reactor R&D	X			X				X																X							
Nuclear stimulation of gas from tight formations																															
Advanced converters & alternative breeder technologies			X												X														X		
Nuclear safeguards (from materials diversion) R&D			X					X							X									X					X		
Storage of commercial nuclear wastes R&D								X							X														X		
Nuclear fuel reprocessing R&D	X		X	X				X							X	X													X		
Fusion research	X		X	X				X							X	X													X		
SO2 control technologies	X		X	X				X							X														X		
NOx, particulate & trace element control R&D	X		X	X				X							X																
Thermal pollution control	X		X	X				X							X																
Conversion of wastes R&D	X		X	X				X							X														X		
Coal gasification	X		X	X				X							X														X		
Coal liquefaction	X		X	X				X							X														X		
Coal mine health and safety research	X							X							X																
Improve coal combustion								X							X																
Coal extraction & reclamation R&D								X																							
Coal cleaning (precombustion) R&D			X					X																							
Advanced coal combustion, including fluidized bed R&D								X																							
Advanced energy conversion R&D								X																							
Environmental effects of fossil fuel utilization R&D								X							X																
Magnetohydrodynamic power cycles	X		X																										X		
Underground electric transmission R&D	X																														
Electric transmission & distribution R&D			X					X							X														X		
FPC & states allow utility recovery of R&D expenses	X		X																										X		
Solar energy R&D	X		X					X							X									X					X		
Initiation of Solar Energy Research Institute (SERI)																													X		
Geothermal energy R&D			X																					X					X		
Geothermal resource assessment & R&D								X							X														X		
Energy conservation R&D			X					X							X														X		
Advanced auto power systems			X					X																					X		
Energy storage R&D								X							X														X		
Oil and gas recovery R&D								X							X														X		
Synthetic fuels "pioneer" program																								X					X		
Aircraft fuel economy R&D (NASA)																													X		
Leasing Federal Lands																															
Accelerate oil and gas leasing on Outer Continental Shelf (OCS)	X														X							X		X					X		
Repurchase oil and gas leases in Santa Barbara Channel	X																														
Oil shale leasing	X		X																										X		
Geothermal leasing	X		X																										X		

Item	1	2	3	4	5	6	7	8	9	10	11	12	13	14	15	16	17	18	19	20
Accelerate development on oil shale and geothermal leases															X		X			
Resume leasing of federal coal lands w/environmental safeguards															X		X			X
Diligence requirements for developing leased coal lands																	X			

Energy Production and Transportation

Item	1	2	3	4	5	6	7	8	9	10	11	12	13	14	15	16	17	18	19	20
Modernize & expand uranium enrichment capacity	X		X																	
Standardize nuclear plant designs			X																	
Speed up nuclear plant licensing			X					X				X			X					
Temporary operating licenses for nuclear plants without public hearing								X												
Defer commercialization of chemical reprocessing of nuclear fuel																				X
Maintain U.S. role as supplier of nuclear reactors and enrichment services																				X
Speed up program to demonstrate commercial high-level nuclear waste storage																				X
Authorize ERDA agreements with pvt. firms to build uranium enrichment plants																				X
Authorize new pricing for ERDA enrichment services																				X
Uranium resources assessment & information																				X
Facilitate LNG imports	X																			X
Energy Resources Council evaluation of LNG imports																				X
Energy facilities siting legislation (ease permitting process)	X	X	X	X				X	X			X	X			X		X		X
Mined Areas Protection Act (strip mining)		X	X	X		X		X	X			X	X			X				
Avoid excessive restrictions on surface mining of coal																				X
Coal slurry pipeline right-of-way																				
Alaska oil pipeline right-of-way			X	X		X		X	X			X				X		X		
Issue permit for construction of trans-Alaska oil pipeline																				
License construction and operation of deepwater ports			X			X		X	X			X								X
Expedited environmental review and selection of route for Alaskan gas pipeline																				
Complete enviro. & economic analyses of Alaskan gas pipeline to lower 48																X				
Mineral leasing act amendments			X						X			X	X							
BLM Organic Act amendments			X						X											
Explore and/or develop Elk Hills Naval Petroleum Reserve, Alaska NPR#4						X		X	X			X	X			X	X	X		X
Evaluation of incentives & regulations to stimulate synthetic fuels production												X								
Use Defense Production Act authority to allocate materials for energy projects																X				
National synthetic fuels commercialization program																	X			X
Energy development area impact assistance																				X
Energy Independence Authority-Gov't corp. to provide financing for energy projects																				X
Geothermal facility loan guarantees																				X

Environmental Protection

Item	1	2	3	4	5	6	7	8	9	10	11	12	13	14	15	16	17	18	19	20
EIS for Breeder program.	X																			
Increased inspection for OCS production			X																	
States delay implementation of secondary Clean Air Act standards			X																	
Study environmental impact of oil & gas production; Atlantic OCS, Alaska			X	X																
State incentives (return) for utility environmental controls			X																	
OMB study conflicts between energy & environmental goals					X															
Temporary relaxation of Clean Air regulations						X														
Legislation to authorize temporary relaxation of Clean Air Act requirements												X								
Allow continued use of coal in power plants if primary standards met																X				
Amend Clean Air Act to allow greater use of coal in electric generation																	X		X	X
Legislative action to "clarify" court significant deterioration decision																	X		X	X
Extend compliance date for scrubbers; allow intermittent controls																	X		X	X
Five-year pause in tightening auto emission standards																	X		X	X
Oil spill liability legislation																				X

Columns grouped by administration — **President Nixon**: 1971 and 1973–1974 dates; **President Ford**: 1974–1976 dates. Header dates shown as Year / Month / Day.

Proposal or Action:	71 6/4	73 2/25	73 4/18	73 6/29	73 8/9	73 9/8	73 10/2	73 10/9	73 10/11	73 10/12	73 11/8	73 11/25	73 12/4	73 12/19	74 1/15	74 1/23	74 3/6	74 4/19	74 6/14	74 6/25	74 7/2	F74 10/8	F74 10/11	F75 1/15	F75 5/27	F75 7/14	F75 12/22	F76 2/26	F76 10/28
Oil Storage/Emergency Reserve Supplies																													
Study oil storage, shut-in production to buffer interruption			X																										
Create Strategic Petroleum Reserve																						X			X			X	
Use receipts from Elk Hills to finance Strategic Petroleum Reserve																						X			X			X	
Energy Efficiency & Conservation																													
Public information on efficient energy use	X																												
Develop conservation measures for industry, governments, public	X																												
Report on success of federal & status of business energy conservation efforts																				X									
Interior Conservation Office study add'l energy conservation measures				X																									
Federal Energy Off. to collect & publicize information on energy efficient products																	X												
Commerce Dept. to develop guidelines for efficient energy use by industry																						X							
Energy Resources Council provide info on conservation methods & benefits																								X					
Encourage governors, mayors & general public to reduce gasoline use									X																				
Mass transit & carpool use, 50 MPH limit, bus lanes, etc.									X																				
Encourage governors, mayors to establish energy emergency offices									X																				
Assistance to states for planning & implementing conservation programs (EPCA)																											X		
Mandatory 50 MPH speed limits/55MPH for trucks to be set											X																		
More rigid enforcement of 55 MPH speed limits										X																			
All federal agencies to develop programs to conserve energy									X																				
Federal buildings lower thermostats										X	X																		
Reduce federal vehicle speed to 50 MPH									X																				
Reduce anticipated federal government energy consumption by 7%				X																									
List of 9 specific actions to reduce energy use by federal gov't & employees									X																				
Major exec. branch agencies cut energy use 24% below 1st 6 months of 1973									X																				
Continue federal energy management (conservation) program				X																									
Reduce anticipated energy consumption by 5% nationwide - 12 months; drive slower, reset thermostats, flight speeds, etc.									X																				
New list of ways to encourage less commercial & residential energy use																						X							
Goals and reporting for industrial energy conservation (IAW EPCA)																													
Compile information on life cycle costs	X																												
Bureau of Standards evaluate home energy use	X																												
Insulation standards for FHA insured loans		X	X																										
Insulation standards for mobile homes		X	X																										
Mandatory thermal efficiency standards for new commercial & residential buildings			X																					X					
Voluntary labeling program for appliances			X				X																	X					
Voluntary appliance efficiency standards; mandatory if necessary		X	X				X																		X			X	
Voluntary labeling program for automobiles		X	X																						X			X	
EPA requirements for fuel economy labels on automobiles																									X			X	
Legislation to authorize mandatory labeling of appliances, automobiles																						X				X		X	
Regulations implementing EPCA mandatory appliance labeling																													X
Request auto industry for 5-yr. schedule to produce fuel efficient autos														X															
Increase auto fuel efficiency by 40% by 1980 w/relaxed emission requirements																					X			X					
Mandatory automobile efficiency standards		X	X																								X		X
FAA to encourage airlines to conserve fuel											X																		

Allocate Highway Trust funds for mass transit
Priority for DOT grants for bus purchases
Authorize greater use of Highway Act aid funds for mass transit
Legislation for mandatory conservation (gaslights, speed, hours)
Outdoor advertising & decorative lighting to be prohibited
Legislation to authorize mandatory conservation and rationing
Legislation to authorize year-round daylight savings time
Legislation to authorize 1 year case-by-case waivers of environmental requirements
Authorize limited exemptions for environmental impact statements
Veto of "Energy Emergency Act" "low interest loans"
Regulatory agencies consider energy use in determinations
Request utility regulators to change rates that encourage energy use
Direct subsidies to low income & elderly for energy conservation
Develop standby conservation regs in event of another embargo (IAW EPCA)

Fuel Switching

Encourage utilities to convert from residual oil to coal
Eliminate use of oil & natural gas in electric generation where feasible
Prohibit utility & industrial facilities from switching from coal to oil
Prohibit powerplants in planning stage from using oil and gas/coal conversion
FEA issued (or would by 6/30/75) orders for converting 31 powerplants to coal

Energy Allocation & Price Controls

Remove restrictions on Canadian oil imports
Limit oil imports, using presidential power if necessary
Deregulate natural gas wellhead prices
Authorize FPC to suspend wellhead price controls during emergency
Allow temporary purchase of uncommitted intrastate gas at unregulated prices
Voluntary oil allocation program
Government royalty oil allocated to small refiners
Proposed mandatory oil allocation program
Mandatory Propane Allocation Program
Mandatory Middle Distillate Allocation Program
Consider limits on fuel blending (to conserve distillates)
Allocation of jet fuel
Encourage governors & mayors to work with federal allocators
Oil allocation legislation
Divert crude from gasoline to heating oil
Reduce refiners' distribution of gasoline to wholesalers, retailers (15%)
Phased reduction of jet fuel by added 15% (to 25%)
Mandatory middle distillates allocation extended to end users
Initiate gasoline allocation program
Reduce fuel available for general aviation
Divert electricity from AEC uranium enrichment facilities when needed
Develop standby rationing program
Legislation to authorize rationing based solely on energy considerations
Standby authority for allocation, rationing
Develop standby allocation regs in event of another embargo (IAW EPCA)
Close gas stations 9 p.m. Saturday until midnight Sunday
Veto of "Energy Emergency Act" containing rigid oil price controls
Use FEA's old oil entitlements program to spread impact of oil import fee
Stop coverage of oil product imports in entitlements program; adjust rebates
Remove price controls on domestic crude by 4/1/75

PROPOSALS AND ACTIONS LISTED IN PRESIDENTIAL ENERGY MESSAGES AND STATEMENTS - Nixon and Ford Administrations (continued)

Proposal or Action:	71 6/4	73 2/25	73 4/18	73 6/29	73 8/8	73 9/8	73 10/2	73 10/9	73 10/11	73 11/12	73 11/8	73 11/25	73 12/4	73 12/19	74 1/15	74 1/23	74 3/6	74 4/19	74 6/14	74 6/25	74 7/2	F 74 10/8	F 74 10/11	F 75 1/15	F 75 1/15	F 75 5/27	F 75 7/14	F 75 12/22	F 76 2/26	F 76 10/28
Energy Allocation & Price Controls (continued)																														
Reduce disincentive in price controls for secondary/tertiary oil recovery																						X								
Directed FEA to decontrol "old" oil in two years																														
Compromise plan to phase out price controls on domestic oil by 1/78, price ceiling on all domestic crude; threaten veto of price control authority																											X			
Oil price formula providing for decontrol after 40 months																												X		
Taxes and Fees																														
SO2 emission charge	X	X																												
Consider "control fees" to reduce gas & electric use			X	X																										
Restructure oil import limitations; reduce fees			X	X																										
Remove tariffs on imported crude oil & products																														
Impose $2/barrel excise tax on domestic crude & $2 fee on all oil imports											X																			
Oil import fee of $3 per barrel (starting at $1, growing to $3 over 3 months)																						X		X						
Imposed 2nd $1 import fee on crude and $.60 on products after president delayed action for 90 days at Congress' request																								X						
Eliminate foreign depletion allowance; modify system of foreign tax credits																														
Impose excise tax on natural gas of $.37 per thousand cubic feet																														
Economic Stabilization Act may be used to avoid excess profits											X																			
Establish windfall profits tax on crude oil																		X												
Windfall profits tax on domestic crude to offset proposed price decontrol																		X												
Broad authority for president to impose oil tariffs, import quotas, price floors																	X													
Windfall profits tax on domes. crude (w/plowback) to offset price decontrol																								X		X				
Tax credit for residential energy saving investments																								X		X				
Extend investment tax credit for exploratory drilling																		X												
Increase investment tax credit for utilities from 4% to 7% available for others																						X		X		X				
Temporary 2-yr. increase in investment tax credit for coal, nuclear powerplants																								X		X				
Allow deduction of preferred stock dividends to reduce costs of capital																								X		X				
Electric utility construction incentives: 12% ITC, 5-yr. amortization for pollution control expenditures & converting or replacing oil-fired; defer taxes on dividends															X									X		X				
International Activities																														
Work with OECD nations on sharing oil in emergency			X	X																										
Work on international cooperation in energy R&D			X	X																										
Participation in new International Energy Agency (consumer nations)																						X		X						
Standby authorities for president to implement international energy agreements: conservation, allocation, rationing, production, etc.																						X		X						
Measures to deal with international financial disruption resulting from financial outflow for oil imports																								X						
Create International Energy Institute to assist developing countries																														
Call upon all nations to avoid transferring nuclear reprocessing & enrichment technologies & facilities																														X
Assure adequate nuclear fuel supplies to countries forgoing building of enrichment and reprocessing facilities																										X				X
Urge all nations to join in strengthening proliferation controls																														X
Cut off nuclear cooperation w/any nation violating proliferation agreement																														X
Tighten proliferation controls on U.S. nuclear exports																														X

State Public Utility Commission Actions

- Require rate decisions within 5 months, fuel adjustment pass-through
- including CWIP in rate base, permitting lower rates during off-peak hours,
- and allowing pollution control equipment costs in rate base
- Energy Resources Council to study need for additional utility regulatory changes
- FEA funding for state & local studies of utility rate structures

Organizational Changes

- Consolidate energy in a Dept. of Natural Resources
- Create a Dept. of Natural Resources (revised proposal)
- Create Office of Energy Conservation in Interior Dept.
- Create Office of Energy Data & Analysis in Interior
- Create special energy committee in White House
- Appoint consultant to president for energy, w/staff (National Energy Office)
- Abolish National Energy Office and special committee
- Appointed Counsellor to President for natural resources, environment, energy
- New Interior Asst Sec. for energy & minerals
- Placed responsibility for oil policy committee in Treasury Dept.
- New Energy Policy Office in White House (Gov. Love)
- Energy R&D Advisory Council in EOP (Wineberg) reporting EPO (Love)
- Abolish White House Energy Policy Office
- Create Energy R&D Administration (ERDA)
- Activate ERDA
- Create Division of Energy & Science in OMB
- Create Nuclear Energy Commission (AEC reg functions; became NRC)
- Creation of Nuclear Regulatory Commission (NRC)
- Study organization of energy regulatory functions (OMB)
- OMB to establish task force to monitor allocation & rationing
- Establish fuel allocation administration in Interior Dept.
- Commerce Dept. establish National Industrial Energy Conservation Council
- Interior to activate Emergency Petroleum Supply Committee
- Create Energy Emergency Action Group
- Abolish Energy Emergency Action Group
- Activate Emergency Petroleum & Gas Executive Reserve
- Legislation to create the Federal Energy Administration (FEA)
- Exec. Order effectuating FEA Act
- Create Fed. Energy Office in EOP (replace EPO), pending legis. creating FEA
- Abolish Federal Energy Office (replaced by statutory FEA)
- Create cabinet level Committee on Energy (Simon Chair)
- Abolish Oil Policy Committee
- Creation of Energy Resources Council in EOP (in act creating ERDA & NRC)

Miscellaneous

- Job security assistance in areas with large increases in unemployment
- Veto of "Energy Emergency Act" - "energy related" unemployment payments
- Mandatory reporting of energy information
- GSA to consider energy implications in land & building acquis. & disp.;
- Fed. Prop. Council to consider surplus property for energy sites

March 6, 1974: President's message vetoing the "Energy Emergency Act"

April 19, 1974: Executive order abolishing the Energy Policy Office

June 14, 1974: Presidential announcement—creation of a cabinet-level Committee on Energy

June 25, 1974: Executive order—effectuation of the Federal Energy Administration Act

July 2, 1974: Statement by the president on energy conservation

President Gerald Ford

October 8, 1974: Address to joint session of Congress, with program to control inflation

October 11, 1974: Statement on signing act to create Energy Research and Development Administration (ERDA), Nuclear Regulatory Commission, and Energy Resources Council

January 15, 1975: State of the Union message (and fact sheet)

January 15, 1975: Presidential statement with executive order activating ERDA

May 27, 1975: Fact sheet on president's administrative actions on energy

July 14, 1975: President's Compromise Oil Decontrol Plan

December 22, 1975: President's signing of Energy Policy and Conservation Act containing some Ford energy proposals

February 26, 1976: President's energy message

October 28, 1976: President's statement on nuclear policy, including nuclear wastes

Not Included in Table

President Ford's January 30, 1975, letter transmitting his proposed omnibus energy bill, which followed up on his January 15, 1975, State of the Union message

President Ford's March 4, 1975, veto of H.R. 1767, Suspension of Oil Import Tariff Authority, which would have restricted the president's authority to impose import fees

President Ford's May 20, 1975, veto of the Surface Mining Control and Reclamation Act

Various statements issued at the time of signing some energy legislation

Appendix B. Technical Assumptions Used in the Scenarios of World Energy Demand

The following are the technical assumptions underlying the energy demand scenarios discussed in chapter 11.

Per Capita Income Growth

The average per capita income growth rates (percent per year) used in the estimates are as follows (figures rounded):
—OECD, 2.0 percent
—Sub-Saharan Africa, 0.5 percent rising to 1.0 percent
—East Asia and Pacific, 5.0 percent
—South Asia, 2.7 percent
—Latin America and Caribbean, 2.0 percent rising to 2.5 percent

In retrospect, rather higher growth rate assumptions should have been considered for Africa for the scenarios being discussed; this would, of course, imply slightly greater energy demands than are shown in figures 11.3 and 11.5 in the text.

For Eastern Europe and Central Asia, the energy demand model was not used to assess energy growth. We deferred to the analysis of the International Energy Agency (OECD/IEA, 1993), which points out that the economic and political changes in these regions do not presently lend themselves to the sort of energy demand modeling analysis used in the OECD and developing countries. The IEA projections show primary energy demands declining from about 1.67 billion tons of oil equivalent (Btoe) in 1990 to 1.37 Btoe in 2000, and recovering to 1990 levels in the first decade of the next century.

Income Elasticities of Demand

The income-elasticity values presented in table B.1 are aggregate estimates for developing ("LDC") and OECD countries. Those for the OECD countries are based on the elasticities implicit in the findings of the IEA's 1993 *World Energy Outlook* (OECD/IEA, 1993), and those for the developing countries were supplied by the World Bank's International Economics Department.[1]

Own Price Elasticities of Demand

The primary source for price elasticities is Bates and Moore (1992). This source contains estimates of elasticities for several developing countries as well as aggregated values for OECD countries as a group. In this analysis these results are used to make assumptions about price elasticities; the values used are shown in table B.2.

The figures shown are (numerically) larger than the estimated elasticities for some regions. In some developing countries, price elasticity estimates for petroleum, gas, and electricity are as low as −0.3. In order not to understate the effects of price changes on demand, we have chosen to round the estimates to −0.5. As indicated, the elasticities we have used are all long-run price elasticities; this is partly because we are interested in long-run effects, and also because they lead us to overestimate rather than underestimate the effects of price changes on demands. This suits the purpose of the present analysis.

Carbon Taxes

In order to estimate the effects of carbon taxes on the demands for the various fuel types, it is necessary to estimate how they would impinge on different fuels. Their impact on coal prices per unit of energy delivered would of course be higher than their impact on the prices for oil and gas. Table B.3 shows the estimated effects on the prices of solid, liquid, and gaseous fuels relative to their average bulk prices, including excise taxes, in the OECD economies. The average prices are weighted by the quantities consumed in the various countries. In the analysis, the taxes are assumed to be phased in linearly between 1995 and 2010.

Price Distortions in Developing Countries

Actual prices in developing countries have changed significantly in recent years with political and economic changes, and there is no readily available up-to-date source of information. Table B.4 (from Bates and Moore, 1992) contains the best judgments we could use as a working basis.

Table B.1.   Average Income Elasticities for Developing and OECD Countries

	Petroleum	Gas	Coal	Electricity
LDCs	1.0	1.5	0.7	2.0
OECD	0.3	1.1	0.4	1.0

Table B.2.   Average Long-Run Price Elasticities for Developing and OECD Countries

	Petroleum	Gas	Coal	Electricity
LDCs	−0.50	−0.50	−0.50	−0.50
OECD	−0.45	−0.70	−1.30	−0.45

Table B.3.   Average Market Prices for Liquid, Gaseous, and Solid Fuels in the OECD and the Effect of Imposing a $100 or $300 Carbon Tax

Energy Source	Average Bulk Price per Ton of Oil Equivalent (1990 U.S. $)		
	Late 1980s	Post $100 Carbon Tax	Post $300 Carbon Tax
Liquids	443	526	695
Gaseous	212	276	405
Solids	63	170	385

Table B.4.   Estimated Energy Price Distortions in Developing Countries

Energy Source	Current Market Prices as a Percentage of Average Incremental Costs of Supply
Coal	50–70
Liquids	80
Gas	30–50
Electricity	60–70

Average incremental costs of supply were estimated by Bates and Moore using border prices for the three fossil fuels and from data on marginal costs in the case of electricity supply. As with carbon taxes in OECD countries, the price reforms are phased in gradually over the period 1995–2010.

NOTE

1. I am grateful to Mudasser Imran for his help and advice on price and income elasticity assumptions.

REFERENCES

Bates, R., and E. Moore. 1992. *Commercial Energy Efficiency and the Environment.* World Bank Policy Research Working Paper. Washington, D.C.: World Bank.

Organization for Economic Cooperation and Development, International Energy Agency (OECD/IEA). 1993. *World Energy Outlook.* Paris: OECD.

Abbreviations

AEC	U.S. Atomic Energy Commission (1946–73)
AFVs	alternatively fueled vehicles
API	American Petroleum Institute
Btoe	billion tons of oil equivalent
CAFE	corporate average fuel economy
CEA	Council of Economic Advisers
CGIAR	Consultative Group on International Agricultural Research
CO_2	carbon dioxide
CRADA	cooperative research and development agreement
DOE	U.S. Department of Energy
DSM	demand-side management
EERC	Energy, Environment, and Resources Center (University of Tennessee)
EIA	Energy Information Administration
EPA	U.S. Environmental Protection Agency
EPCA	Energy Policy and Conservation Act (1975)
ERDA	U.S. Energy Research and Development Administration (1974–77)
FEA	Federal Energy Administration (1973–77)
FERC	Federal Energy Regulatory Commission
GATT	General Agreement on Tariffs and Trade
GDP	gross domestic product
GEF	Global Environment Facility
GNP	gross national product
IEA	International Energy Agency; a part of OECD
IPP	independent power producers
IRP	integrated resource planning
JIEE	Joint Institute for Energy and Environment
kW	kilowatt(s) (equal to 1 thousand watts)
kWh	kilowatt-hours

kWp	peak kilowatt capacity
LDC	less developed country
MPG	miles per gallon
MTBE	methyl tertiary butyl ether (an octane enhancer in motor fuels)
MW	megawatt(s) (equal to 1 million watts)
NOx	nitrogen oxides
NRC	U.S. Nuclear Regulatory Commission
OECD	Organization for Economic Cooperation and Development
OMB	Office of Management and Budget
OPEC	Organization of Petroleum Exporting Countries
ORNL	Oak Ridge National Laboratory
OSP	official selling price
PM	particulate matter
PURPA	Public Utility Regulatory Policies Act (1978)
PV	photovoltaics
R&D	research and development
RD&D	research, development, and demonstration
RFF	Resources for the Future
SO_2	sulfur dioxide
SPR	Strategic Petroleum Reserve
TAP	trans-Alaska pipeline
TVA	Tennessee Valley Authority
TW	terawatt
UNDP	United Nations Development Program
UNEP	United Nations Environment Program
VOC	volatile organic compound

Contributors

DENNIS ANDERSON is an adviser on energy and industry at the World Bank. He was the principal economist for the bank's *World Development Report 1992: Development and the Environment.* He has held several research and operations positions at the World Bank as a senior economist, in practically all sectors of bank activity. Before joining the World Bank he was an engineer in U.K. industry. He was also chief economist at Shell from 1987 to 1989, a visiting professor at University College London, and, on three occasions, a visiting scholar at Oxford University. He has published widely on energy, growth, and development and has been involved in the preparation of several World Bank policy papers. He is the co-editor of the *Annual Review of Energy and the Environment.*

JOHN BERRY covers the economy and economic policy issues in Washington, D.C. In 1969 he became a correspondent in the McGraw-Hill World News Service's Washington bureau, covering economics and writing for *Business Week.* Four years later he joined *Time* as its national economics correspondent. In 1977 he moved to *Forbes* as its Washington bureau manager. For the last fifteen years he has covered economics for the *Washington Post.* His stories are carried by the *Los Angeles Times–Washington Post* News Service. He is also a frequent contributor to the *International Herald-Tribune,* and his work has appeared in *Fortune, Financier,* and other American and foreign publications.

DOUGLAS R. BOHI is director of the Energy and Natural Resources Division at Resources for the Future, a nonprofit research institution located in Washington, D.C., that specializes in resource and environmental issues. During 1987–88 he was on leave from RFF as chief economist and director of the Office of Economic Policy at the Federal Energy Regulatory Commission. One of his principal responsibilities was developing recommendations for reforming FERC regulation of the wholesale power market. Bohi is the author or co-author of seven books on energy issues and has contributed many articles to professional journals and books. In addition to teaching at Southern Illinois University and working as an economist at Caterpillar Tractor Company and the Office of the Assistant Secretary of Defense for Systems Analysis, he has been a Fulbright Scholar in the Netherlands and a visiting professor at Monash University in Melbourne, Australia.

ROBERT A. BOHM is a professor of economics at the University of Tennessee–Knoxville. He serves as director of graduate studies in the Economics Department and as associate director of the university's Energy, Environment, and Resources Center. He specializes in energy and environmental economics, public finance, and urban and regional economics. He has directed or co-directed twenty-six research projects and written widely in these areas. His most recent work has concentrated on determining the need for and financing of hazardous- and solid-waste facilities. He is currently completing a study of environmental taxation. During his career Bohm has worked abroad in the Philippines, Nicaragua, Germany, the United Kingdom, and China on projects such as the development of potable-water systems in rural areas, the reclamation of lands following surface coal mining, and the design of appropriate pollution effluent fees. He has served as a member of the Board of Directors of the National Tax Association and as a consultant to the U.S. Army Corps of Engineers, the Tennessee Valley Authority, Oak Ridge National Laboratory, Argonne National Laboratory, and the U.S. Agency for International Development.

MICHAEL E. CANES became vice president of Policy Analysis, Statistics, and Information Systems at the American Petroleum Institute in 1982, after having served as senior economist, deputy director, and then director of the API's Policy Analysis Department. Before joining the API in 1974 he was an assistant professor at the University of Rochester's Graduate School of Management and an economist at the Center for Naval Analyses. His writings have appeared in the *Oil and Gas Tax Quarterly,* the *Journal of Energy and Development,* the *Journal of Business,* the *American Economic Review,* the *Middle East Review,* and *Applied Economics.* He has presented papers at numerous conferences sponsored by the World Energy Council, the U.S. Department of Energy, the American Enterprise Institute, the Western Economics Association, and the American Institute of Mining, Metallurgical, and Petroleum Engineers.

JOEL DARMSTADTER has been with Resources for the Future in Washington, D.C., since 1966. He is presently a senior fellow in RFF's Energy and Natural Resources Division, specializing in economic aspects of energy and the environment. Prior to joining RFF, he was an economist with the National Planning Association. He has served on a number of National Research Council panels and federal government advisory committees. Since 1983 he has been a professional lecturer at the Johns Hopkins University School of Advanced International Studies. He is author or co-author of five books and numerous journal articles, and—jointly with Michael Toman—he edited and contributed to the 1993 RFF study, *Assessing Surprises and Nonlinearities in Greenhouse Warming.*

A. DENNY ELLERMAN is executive director of the Center for Energy and Environmental Policy Research and a senior lecturer at the Sloan School of Management, Massachusetts Institute of Technology. He was formerly vice president of Charles River Associates, executive vice president of the National Coal Association, and deputy assistant secretary for policy analysis, U.S Department of Energy. He has also served on the staffs of the National Security Council, the U.S. Energy Research and Development Administration, and the Office of Management and Budget. He was elected president of the International Association for Energy Economics for 1990. His CEEPR-supported research focuses on productivity change in the coal industry and its effect on coal prices, as well as the relationships between coal and emission allowance prices.

DAVID LEWIS FELDMAN is a senior research associate at the Energy, Environment, and Resources Center, University of Tennessee–Knoxville and an adjunct professor of political science. His research interests include public involvement in land use and natural resource decisions, including water policy, international activities to address global environmental problems, and radioactive waste policy and management. His publications have appeared in such journals as *Policy Studies Journal, Policy Sciences, Global Environmental Change, Society and Natural Resources,* the *Journal of Public Policy, Environmental Professional, Policy Currents, Political Psychology,* and the *World Resources Review* and in several books. He is the author of *Water Resources Management: In Search of an Environmental Ethic* and *Global Climate Change and Public Policy.* He serves on the board of editors of the *Policy Studies Journal,* and is a research team leader for the National Center for Environmental Decision-Making Research, an NSF-funded center at UT.

CHRISTOPHER FLAVIN is vice president for research at Worldwatch Institute, a nonprofit organization in Washington, D.C., devoted to the analysis of global resource and environmental issues. He is a regular co-author of the annual *State of the World* report published in all the world's major languages. His research focuses on energy resource, technology, and policy issues. He has authored several Worldwatch papers and is the co-author of three books, including *Renewable Energy: The Power to Choose* and *Saving the Planet.* He has testified before U.S. House and Senate committees and has appeared at parliamentary hearings in Canada, Italy, and the United Kingdom. His advice has also been sought by major corporations, including Royal Dutch Shell, Enron Corp., and General Motors. He has published articles in over fifty popular and scholarly periodicals. In 1990 Flavin served as an adviser on sustainable development to the government of Bhutan. He participated in the United Nations Earth Summit in Rio de Janeiro in 1992 and helped draft energy position papers for the Clinton-Gore campaign. Also in 1992 he helped found and now serves on the board of the Washington-based Business Council for a Sustainable Energy Future.

PETER S. FOX-PENNER is principal deputy assistant secretary for energy efficiency and renewable energy at the U.S. Department of Energy. He is responsible for directing the planning, implementation, and management of major projects and programs in the Office of Energy Efficiency and Renewable Energy. During the 1992 presidential campaign he served as an energy and environmental adviser on a wide variety of issues. In addition to speaking engagements and briefings, he was active in Environmentalists for Clinton-Gore. He was formerly a vice president of Charles Rivers Associates, Inc. A specialist in regulated industries and energy and environmental policy, Fox-Penner has advised numerous public and private clients and has published a number of articles and books. He has provided expert-witness testimony before the Interstate Commerce Commission, federal courts, the Federal Energy Regulatory Commission, and a number of state public service commissions. From 1980 to 1983 he served as chief engineer in the Illinois Governor's Office of Consumer Services. Prior to this he worked for the University of Illinois Energy Research Group. During this period he was a founder and leader of the Prairie Alliance, a grassroots environmental group.

DAVID L. GREENE is a senior research staff member at Oak Ridge National Laboratory's Center for Transportation Analysis. After joining ORNL in 1977, he founded the Transportation Energy Group (1980) and the Transportation Research Section (1987). He spent 1988–89 in Washington, D.C., as a senior research analyst conducting research on automotive fuel economy and alternative transportation fuels policy in the Office of Domestic and International Energy Policy, U.S. Department of Energy. He has written fifty articles for professional journals and an equal number of technical reports. Active in the Transportation Research Board and the National Research Council, he recently served on the NRC's Committee on Automobile and Light Truck Fuel Economy and is past chairman of the TRB's Section on Environmental and Energy Concerns.

WILLIAM W. HOGAN is Thornton Bradshaw Professor of Public Policy and Management at the John F. Kennedy School of Government, Harvard University. He is the former chairman of the Public Policy Program at the Kennedy School and a former director of its Energy and Environmental Policy Center. As director of the Project on Economic Reform in Ukraine, he is providing assistance on the broader problems of economic transition to a market economy. He has served on the faculty of Stanford University, where he founded the Energy Modeling Forum, and is a past president of the International Association for Energy Economics. He has held positions dealing with energy policy analysis in the Federal Energy Administration, including that of deputy assistant administrator for data and analysis. He is a director of Putnam, Hayes, and Bartlett, Inc., and is involved in various research and consulting activities including major energy industry restructuring, network pricing and access issues, and privatization in several countries.

GEORGE HORWICH is a professor of economics at Purdue University, where he has taught since 1956. He has also taught at Indiana University, the University of California at San Diego, and the People's University of China. His early publications were in monetary and macroeconomic theory and included *Money, Capital, and Prices* (1964) and *Trade, Stability, and Macroeconomics* (1974), co-edited with Paul Samuelson. In 1978–80 and 1984 he was a senior economist with the Department of Energy's policy office, following which he published *Oil Price Shocks, Market Response, and Contingency Planning* (1984), co-authored with David Weimer, and *Responding to International Oil Crises* (1988), co-edited with David Weimer. More recently he has written on the economics of natural disasters, the economics of mobilization, and the economy of China. He is an adjunct scholar of the American Enterprise Institute and has served on the staffs of the National Bureau of Economic Research, the Brookings Institution, and, as a collaborating scientist, the Oak Ridge National Laboratory. He has held several named professorships at Purdue.

KATHRYN J. JACKSON is currently senior vice president of the Resources Group at the Tennessee Valley Authority. Her organization has responsibility for developing and demonstrating technologies in support of TVA and the Tennessee Valley. Previously she was a project engineer at TVA, helping the agency examine options for next-generation nuclear power. She completed a fellowship at the National Academy of Engineering in Washington, D.C., where she directed a program examining the allegations that U.S. managers have short time horizons and the implications this may have for long-term technology development. Her thesis in engineering and public policy from Carnegie Mellon University examined the engineering portion of the product development process, with a special emphasis on determining process and product quality. Previously she worked in industry for eight years, for both Westinghouse Electric Corporation and Alcoa, in various positions including design, project marketing, and technology forecasting.

RICHARD D. MORGENSTERN is a visiting fellow at Resources for the Future. He is the former director of the Office of Policy Analysis at the U.S. Environmental Protection Agency's Office of Policy, Planning, and Evaluation. Other previous positions include acting assistant administrator for policy, planning, and evaluation at EPA; acting deputy administrator, EPA; director, Office of Policy Analysis, EPA; director, Energy Program, Urban Institute; legislative assistant, Senator J. Bennett Johnston; deputy assistant administrator for energy, natural resources and the environment, Congressional Budget Office; and associate professor of economics, Queens College of the City University of New York. He has had numerous articles published in peer-reviewed and popular journals and has given extensive congressional testimony. He is a member of the U.S. Team for the Intergovernmental Panel on Climate Change and the International Negotiating Committee. He is the former chairman of the Group of Economics Experts, OECD.

VERRILL M. NORWOOD III has been vice president of technology advancements for the Tennessee Valley Authority since December 1994. He is responsible for a division that identifies, develops, and documents the benefits from research and technology development activities and transfers technologies to improve the performance and competitive health of TVA. He worked for the Pfizer Corporation between 1985 and 1987 as a research engineer. In 1987 he accepted the position of research chemist at TVA's Environmental Research Center in Muscle Shoals, Alabama. During this time he developed new and improved technologies to remediate various waste streams from the production of agricultural chemicals and munitions. In April 1992 he joined the Research and Development organization in TVA's Chattanooga offices as manager of environmental control technology. As a result of the refocusing of the Resource Group, he became manager of environmental technology development in technology advancements in April 1993. His responsibilities include development and demonstration of clean coal technologies, waste reduction and by-product utilization technologies, wastewater treatment technologies, and air emissions control technologies.

ALLAN G. PULSIPHER is director of the Policy Analysis Program at the Center for Energy Studies and a professor in the Institute of Environmental Studies, both at Louisiana State University in Baton Rouge. Previously he was the chief economist for the Tennessee Valley Authority. He has also been the chief economist for the Congressional Monitored Retrievable Storage Review Commission; a program officer with the Ford Foundation's Division of Resources and the Environment; a senior staff economist with the president's Council of Economic Advisers; and a member of the faculties of Texas A&M and Southern Illinois universities.

GLENN R. SCHLEEDE is president of Energy Market and Policy Analysis, Inc., a consulting practice providing analysis, advice, and assistance to organizations in industry and government on energy and related environmental and economic matters. Prior to forming EMPA, he was vice president of New England Electric

System and president of New England Energy, Inc., with responsibilities for oil and gas exploration, procurement and transportation of fuels for NEES facilities, a coal-shipping venture, and the system's economic planning and budgeting functions. Previously he was executive associate director of the U.S. Office of Management and Budget, senior vice president of the National Coal Association, and associate director (energy and science) of the White House Domestic Council. He also held career service positions in the U.S. Office of Management and Budget and the U.S. Atomic Energy Commission.

CHAUNCEY STARR was founding president and vice chairman of the Electric Power Research Institute, where he is now president emeritus. From 1967 to 1973 he was dean of the UCLA School of Engineering and Applied Science, following a twenty-year industrial career, during which he served as vice president of Rockwell International and president of its Atomics International Division. Following World War II he pioneered in the development of nuclear propulsion for rockets and ramjets; in miniaturizing nuclear reactors for space; and in developing atomic power plants. For his work in the peaceful uses of atomic power, he received the Atomic Energy Commission Award in 1974, the Walter H. Zinn Award in 1979, and the Henry D. Smyth Award in 1983. He has also received a Distinguished Contribution Award from the Society for Risk Analysis in 1984. The United States Energy Association selected him as the 1990 recipient of the United States Energy Award for his long-term contributions to energy and to international understanding. In 1990 he was awarded the National Medal of Technology from the President of the United States for his contribution to engineering and the electrical industry, and in 1992 he received the Rene Dubos Environmental Award. Starr is a member of the National Academy of Engineering; a founder and past president of the American Nuclear Society; and a member and past director of the American Association for the Advancement of Science.

Index

43, 49–50; price increases, 64, 81–82; deficits, 94; in OECD countries, 110
Gross national product (GNP), 180; future crises, 12; policy, 14; Project Independence, 116; OPEC, 151, 153; loss of potential, 152; U.S. oil expenditures, 165–67

Haiti, 215
Harte, John, 59
Harvard University, 119, 259
Heinz, John, 259
Hidy, George, 23
Highway Act, 269
Highway Trust, 269
Hirst, Eric, 43
Hogan, William M.: SPR, 9; technology, 15; markets, 43; energy forecasting/modeling, 101–2, 128–35; economic analysis, 137–41, 148
Holdren, John, 59
Horwich, George, 4–5, 17
House of Representatives, U.S., 33
Houthakker, Hendrik, 77
Hussein, Saddam: imports, 23; oil, 74, 85–86; invasion of Kuwait, 117
Hydrogen, 13; hydroelectric dams, 249; hydropower, 238; hydrogen-powered cars, 241. *See also* Magnetohydrodynamic power
Hythane, as fuel, 13

IBM, 245
Iceland, 34
Imran, Mudasser, 276
Inadvertent Climate Modification (MIT), 31
Independent power producers (IPPs), 252–53
India, 238, 243
Induction heating, 252
Industrialized capitalism, 37
Inflation, 4, 56
Institute for Applied System Analysis, 233
Integrated resource planning (IRP), 250–51
Intergovernmental Panel on Climate Change (IPCC), U.N., 233, 236
Interior Assistant Secretary for Energy and Minerals, 271
Interior Conservation Office, U.S., 268

Interior, Department of, U.S. *See* Department of Interior (DOI)
International Conference on Population and Development, 237
International Economics Department, World Bank, 274
International Energy Agency (IEA), 51, 270, 273–74; oil imports, 34; global oil usage, 97–98; oil demands study, 185–86; renewable energy, 200; energy future, 233
International Energy Institute, 270
International Energy Symposia Series (IESS): 1982 World's Fair, 210–12; energy future, 215
Iran, 73, 91; oil industry, 26; Ayatollah, 73
Iranian Revolution, 43, 85; oil bargaining strength after, 26; oil prices after, 32, 36
Iran-Iraq War, 85, 150
Iraq, 227; seizure of Kuwait, 23–24, 74, 117; oil output, 85, 95–96; nuclear non-proliferation, 219
Israel, 150, 227
Issue: analysis, 210; identification, 210; resolution, 210
"Iterative functionalism," 173

Jackson, Kathryn J., 13, 222–23
Japan: nuclear power, 16, 119; metal imports, 49; oil imports, 50–51; greenhouse gases, 79; energy demand, 178; electricity, 202; industrialization, 244
John F. Kennedy School of Government, Harvard, 259
Johnston, J. Bennett, 74
Joint Congressional Committee on Atomic Energy, 146
"Joint implementation," 175, 208
Jorgenson, Dale W., 120–21, 131

Kahn, Alfred, 262
Kalt, Joel, 83
Kazakhstan, 242
Kennedy, John F., 140
Kenya, 194
Keynesian: economics, 82–83, 257; activism, 257
Khadafi, Moamar, x
Kilometer (km), 198

Library of Congress Cataloging-in-Publication Data

The energy crisis : unresolved issues and enduring legacies / edited by David Lewis Feldman.
  p.   cm.
Includes index.
ISBN 0-8018-5361-3 (alk. paper)
1. Energy policy—United States.   2. Power resources—United States.   3. Petroleum
products—Prices—United States.   4. Energy conservation—United States.   I. Feldman,
David Lewis, 1951–
HD9502.U52E485   1996
333.79'0973—dc20   96-1138